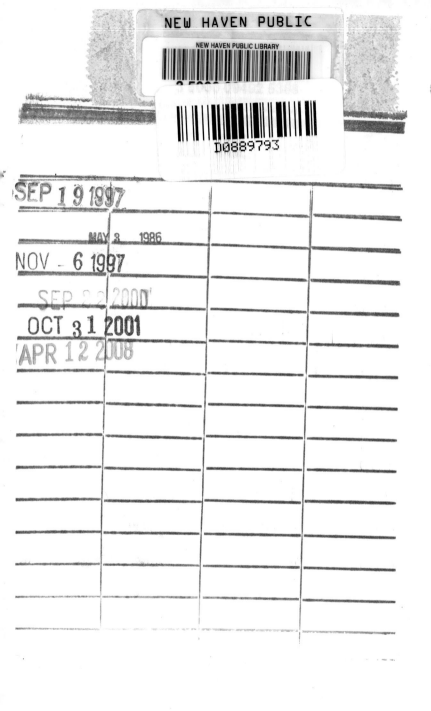

From Knowledge to Wisdom

A REVOLUTION IN THE AIMS
AND METHODS OF SCIENCE

Nicholas Maxwell

Basil Blackwell

© Nicholas Maxwell 1984

First published 1984
Basil Blackwell Publisher Limited
108 Cowley Road, Oxford OX4 1JF, England

Basil Blackwell Inc.
432 Park Avenue South, Suite 1505
New York, NY 10016, USA

British Library Cataloguing in Publication Data

Maxwell, Nicholas
 From knowledge to wisdom.
 1. Objectivity
 I. Title
 121   BD222

 ISBN 0-631-13602-9

10923755

Typeset by Cambrian Typesetters, Aldershot, Hants.
Printed in Great Britain by TJ Press Ltd, Padstow

Contents

In memory of my mother and father
and to Christine van Meeteren

Preface

This book argues for the need to put into practice a profound and comprehensive intellectual revolution, affecting to a greater or lesser extent all branches of scientific and technological research, scholarship and education. This intellectual revolution differs, however, from the now familiar kind of scientific revolution described by Kuhn. It does not involve a radical change in what we take to be knowledge about some aspect of the world, a change of paradigm. Rather it involves a radical change in the fundamental, overall intellectual aims and methods of inquiry. At present inquiry is devoted to the enhancement of knowledge. This needs to be transformed into a kind of rational inquiry having as its basic aim to enhance personal and social wisdom. This new kind of inquiry gives intellectual priority to the personal and social problems we encounter in our lives as we strive to realize what is desirable and of value – problems of knowledge and technology being intellectually subordinate and secondary. For this new kind of inquiry, it is what we do and what we are that ultimately matters: our knowledge is but an aspect of our life and being.

I shall argue that a necessary, though not a sufficient, condition for us to develop cooperatively a better, more humane world is that we have in existence a tradition of rational inquiry of this new kind, giving priority to life and its problems, devoted to the enhancement of wisdom. At present we have no such tradition. As a result we are all more or less severely handicapped in our capacity to resolve in desirable and good ways problems we encounter in our personal and social lives. Many of our present-day social and global problems are in part due to our long-standing failure to develop such a tradition of genuinely rational, socially active thought, devoted to the growth of wisdom. This basic Socratic idea has been betrayed, and as a result, to put it at its most extreme, we now stand on the brink of self-destruction. In the circumstances, there can scarcely be any more urgent task for

all those associated in any way with the academic enterprise –
scientists, technologists, scholars, teachers, administrators,
students, parents, providers of funds – than to help put into
practice the new kind of inquiry, rationally devoted to the growth
of wisdom.

CHAPTER ONE

Human Suffering and the Need for a Comprehensive Intellectual Revolution

Our planet earth carries all too heavy a burden of killing, torture, enslavement, poverty, suffering, peril and death. It has been estimated that over three and a half million people die each year from starvation or from disease related to malnutrition (George, 1976, p. 19). Millions of children suffer from protein deficiency, their brains failing to develop properly as a result. And yet it seems we have the capacity to produce enough food for everyone to get enough to eat, given a more just distribution of land and food, and less wasteful priorities of food production in the developed world. Life expectancy in the developed world is seventy-two years; in the poorer regions of the underdeveloped world it is as low as forty-five years. In the developed world, on average, fewer than two children out of one hundred die during the first year of life; in the poorer regions of the underdeveloped world fifty out of a hundred die during their first year. Somewhere between thirty-five and sixty million people died as a result of the last world war; and a larger number of people have died in wars since then. Dictatorships are commonplace amongst the nations, the criminally insane not infrequently seizing and holding power, dictatorial power being maintained by means of terror, arbitrary imprisonment, torture and execution – and such dictatorships are even supported by democracies. The cold war between east and west continues, together with the nuclear arms race, the balance of terror, and the persistent possibility that nuclear war will before long engulf the world, through escalating bluff and threat, or accident.[1]

[1]The best overall account, to my knowledge, of our human, global problems, and of our present incapacity to respond, sanely and rationally to them, is given by Higgins (1978). Higgins' 'seventh enemy' is the inertia of our institutions, which renders them, and us, incapable of responding to the crisis. The central thesis of the present book is that the intellectual/institutional inertia of the academic enterprise

Danger, suffering and death are inevitable aspects of life, imposed on us as a result of our living in, and being a part of, the natural world. The danger, suffering and death just indicated, experienced by so many, are not however caused solely by natural phenomena: they are our own creation, our own responsibility, caused by our own actions, or by our failure to act.

The problem to be tackled in this book can be put like this. What kind of rational inquiry gives us the best hope of helping us progressively to resolve our most urgent problems of living – such as those indicated above – thus helping us to develop a more humane, a more just, a happier, a saner and more cooperative world? What kind of science, technology, scholarship and education is best designed to help us promote human welfare, realize that which is genuinely of value in life? What ought to be the basic intellectual aims and methods of such an inquiry, and how ought these to be related to our personal and social aims and methods in life?

Insofar as academic inquiry does try to help promote human welfare, it does so, overwhelmingly, at present, by seeking to improve *knowledge* of various aspects of the world. It does this in the hope that new knowledge, thus obtained, will be used to help resolve social problems in a humane and just fashion. The view that rational inquiry ought to help enhance the quality of human life by, in the first instance, improving knowledge is, one might say, the official basic creed of the whole scientific/academic enterprise. The view can be traced back at least to Francis Bacon in the seventeenth century, and perhaps back to the ancient Greeks. It has been almost unthinkingly taken for granted by almost everyone associated with the development of science, scholarship, universities and education in the western world, and elsewhere. And as a result the view is now firmly built into the whole intellectual-institutional structure of the scientific/academic enterprise.

The central claim of this book can now be put like this. Granted that inquiry has as its basic aim to help enhance the quality of human life it is actually profoundly and damagingly *irrational, unrigorous,* for inquiry to give intellectual priority to the task of

is, in a major way, responsible for the general inertia of institutions, social and international relations and arrangements. Other works also to be consulted in order to get some sort of picture of human world-wide problems are Dubos and Ward (1972); Ward (1979); Meadows *et al.* (1974); *Scientific American*, 243, (3); P. Harrison (1979); Foley (1981); Maddox (1972); SIPRI (1979); Allen (1980); Eckholm (1982); The Committee . . . (1981); Goodwin (1982); Schell (1982).

improving knowledge. Rather, intellectual priority needs to be given to the dual tasks of articulating our problems of living, and proposing and criticizing possible solutions, namely possible human actions. Problems of knowledge and understanding need to be tackled as rationally subordinate to intellectually more fundamental problems of living. In order to develop better solutions to the appalling human problems indicated above, it is not primarily new knowledge that we need; rather what we primarily need is to *act* in new, appropriate ways. The fundamental intellectual task of a kind of inquiry that is devoted, in a genuinely rational and rigorous way, to helping us improve the quality of human life, must be to create and make available a rich store of vividly imagined and severely criticized possible actions, so that our capacity to act intelligently and humanely in reality is thereby enhanced. In order to improve our capacity to resolve the appalling problems confronting humanity today, we need, as a matter of urgency, to develop a new more rigorous kind of inquiry, in many ways radically different from what we have at present, having, as its basic aim, to improve not knowledge only, but rather *wisdom*.

There is thus, I claim, a major intellectual disaster at the heart of western science, technology, scholarship and education – at the heart of western thought; and this long-standing intellectual disaster has much to do with the human disasters of our age, our incapacity to tackle more humanely and successfully our present world-wide problems. In order to develop a saner, happier, more just and humane world it is certainly not a *sufficient condition* that we have an influential tradition of rational inquiry devoted to helping us achieve such ends. It is, however, I shall argue, a *necessary condition*. In the absence of such a tradition of thought, rationally devoted to helping us solve our problems of living, we are not likely to resolve these problems very successfully in the real world. It is this which makes it a matter of such profound intellectual, moral and social urgency, for all those in any way concerned with the academic enterprise, to develop a kind of inquiry more rationally devoted to helping us resolve our problems of living than that which we have at present.

In this book, then, I argue for the need to put into practice a profound and comprehensive intellectual revolution affecting to a greater or lesser extent all branches of science, technology, scholarship and education. The intellectual revolution that I advocate differs however from the now familiar kind of scientific revolution so brilliantly described by Kuhn (1962). For I do not here advocate a change in what we take to be knowledge about some aspect of the world, a change of theory or 'paradigm'. Rather, what I advocate is a radical change – a radical evolution – in the overall, fundamental aims and methods of inquiry. At present we have a kind of academic inquiry that has, as its basic intellectual aim, to improve *knowledge*. This needs to be transformed, I shall argue, into a kind of rational inquiry that has, as its basic intellectual aim, to improve wisdom. This new kind of inquiry is, I shall argue, potentially both more rational (more intellectually rigorous) and of greater human value than what we have at present, inquiry restricted to the improvement of knowledge.

I shall develop the argument by articulating, comparing and contrasting two rival views about what the basic intellectual aims and methods of inquiry ought to be. I shall call these two views 'the philosophy of knowledge' and 'the philosophy of wisdom'. In arguing for the need to put into practice the philosophy of wisdom as opposed to what we have at present, inquiry pursued in accordance with the philosophy of knowledge, I shall be arguing for the need to transform our whole conception of the social sciences and humanities: I shall be arguing for the need to develop a new relationship between the social sciences and humanities on the one hand, and the natural sciences on the other hand: above all, I shall be arguing for the need to establish a new relationship between inquiry as a whole and human life, our personal and social worlds. The revolution that I seek to advocate has widespread intellectual repercussions for science and scholarship; it also has repercussions for the whole institutional structure of the academic enterprise, its place and role in human life. The revolution – or evolution – of basic aims and methods for inquiry that I wish to advocate thus combines intellectual and institutional or social changes.

So far, what this book sets out to accomplish has been characterized so briefly, that misunderstandings are more or less inevitable.

Thus, to begin with, it may be thought strange that I should refer to social and political problems that confront us globally, and in the third world, and yet say nothing about problems of the industrially-advanced first world. Is there not poverty, injustice, only partially-realized democracy or even totalitarianism, much unnecessary human suffering and waste and death here too, as well as in the third world? Should not a kind of rational inquiry that is devoted to helping us realize what is of value in life help us to develop more just, cooperative, fruitful ways of life in Europe, the USSR and the USA as well as in Africa, Asia and South America? My answer to these questions is: yes, of course.[2] Above, I merely sought to indicate what seem to me to be the most urgent, the most desperate problems confronting people in the world today. To this I would add, however, that in seeking to change things for the better in the first world, we ought always to take into account the far more severe problems and plight of people in the third world. A good motive for attempting to bring about social and political changes in the industrially-advanced world is indeed just to help the poor of the third world by putting a stop to first-world violence and exploitation in the third world. During the last thirty years or so, the first world east–west conflict has been fought out primarily in the third world – in Korea, Vietnam, Cambodia, Africa, the Middle East, and South and Central America. It is the defenceless poor of the third world who have suffered the worst consequences of first-world conflict so far – either through war, or through USA- or USSR-backing for corrupt, puppet, totalitarian regimes, in Latin America, in Africa, in Asia. Those of us who live in the first world need to strive to put our own house in order in part because of the havoc we help to cause at present elsewhere.

In the second place, puzzlement may arise in connection with my apparently exclusive concern with large-scale social, political, global problems. I declare that we need a new kind of rational inquiry that helps us to realize what is of value – and yet I seem to ignore where it is that all that is really of value in life is to be found. For is it not the case that, for each one of us, what is of value has to do with the particularities of our own personal lives, our experiences, feelings, desires, achievements: what we do and share with those we know and love? Is not salvation always personal and particular, and never to be found in large-scale schemes for the resolution of problems that confront millions of people together? Once again, with this I agree. As the argument unfolds I shall develop and repeatedly emphasize the point that inquiry, if pursued in a genuinely rational fashion (in accordance with the philosophy of wisdom) must be recognized to be fundamentally personal and interpersonal in character, an aspect of life, our own seeking after the realization of what is of value to

[2]On the basis of statistics concerning such things as the incidence of suicide, madness, alcoholism and war, Fromm comes to the tentative conclusion that the sanity of industrially advanced societies must be called into question. (See Fromm 1963.) For a survey of poverty in Britain, see Townsend (1979).

us personally. But I shall also emphasize that inquiry must have a social, institutional and traditional character – even an impersonal aspect – if it is to perform its proper personal function of helping us to get in touch with what is of value in the world, in each other, and in ourselves. In particular, a vital task for inquiry – for education – is to help us to take up our proper share of adult responsibility for our common world. Our most passionate desires, joys and concerns no doubt quite properly have to do with particular people and things of our own personal life: but we ought also to have some care and concern for all those millions and millions of strangers known to us only by hearsay, who also live with us on earth. Only when it is a commonplace for individuals everywhere to have some measure of informed concern for their fellow citizens of the earth, will there be an end to the nightmare dangers and disasters that now beset so many millions of people as a result of actions and attitudes of other people. Perhaps the most important task of the kind of inquiry and education to be characterized and advocated in this book (inquiry pursued in accordance with the philosophy of wisdom) is just to help each one of us individually to inform and enrich our abstract knowledge of millions of strangers with something of what we feel for those few people we are acquainted with, and those even fewer we love – so that we become capable of recognizing the humanity of millions of distant strangers too. Inquiry as at present constituted (pursued in accordance with the philosophy of knowledge) not only fails to help us connect up personal and public realms in this way. Even worse, it actually intensifies the gulf between personal and public worlds, in that it demands, as we shall see, that a decisive gulf be maintained between personal feelings and values on the one hand, and public, objective facts and knowledge on the other hand.

In the third place puzzlement may arise over my apparently exclusive concern with the human, practical or social use and value of science and scholarship. For does not inquiry have an intrinsic intellectual value, quite apart from any practical applications it may lead to? Is not inquiry something worth pursuing for its own sake, a vital part of our culture and civilization in its own right, like poetry and music? Once again, with all this I agree. A major part of the argument of this book is just that the philosophy of knowledge fails to do justice to the intrinsic intellectual value of inquiry, so that pursuing inquiry in accordance with its edicts does much to obscure and sabotage the potential intellectual value of science and scholarship. In order fully to develop and make available the intellectual riches inherent in diverse aspects of science and scholarship it is essential to put the philosophy of wisdom into practice. I shall argue that inquiry pursued for its own sake is, at its best, an aspect of love, our shared endeavour to see, to apprehend that which deserves love, in the world and in ourselves. This is true even of a subject as apparently remote from love as theoretical physics. At its intellectual best, theoretical physics is an expression of our shared love for that aspect of the world which has to do with its underlying structure or architecture. Physics is of intrinsic intellectual value to the extent that it does enable us to see, to apprehend, to love, something of this architectural grandeur inherent in nature. The greatest exponent of physics pursued in this kind of way is perhaps Einstein. It is to this conception and experience of physics that Einstein alluded when he wrote in a letter to a friend: 'You have given me great joy with the little book about Faraday. This man loved mysterious Nature as a lover loves his distant beloved. In his day there did not yet exist the dull specialization that stares with self-conceit through hornrimmed glasses and destroys poetry' (Dukas and Hoffmann, 1979, p. 42). The philosophy of knowledge destroys the poetry of physics not just because it permits dull specialization and self-conceit to flourish. More fundamentally, it does so because it demands that a gulf be maintained between, on the one hand, the intellectual domain of science and knowledge, having to do with objective fact and truth and, on the other hand, the personal domain of 'subjective' experiences, feelings and values, having to do with such things as joy, fear and love. The result is that it becomes nonsensical to speak of physics as a shared act of love for our world. In order to become a lover of the universe, with Kepler, Faraday, Einstein and others, we need to bring together shared concern for objective, impersonal truth and reality, and our own personal instinctive feelings and imaginings. As we shall see, it is just this which the intellectual standards of the philosophy of wisdom encourage and demand. Upholders of the philosophy of knowledge may, or may not, value personal love of 'mysterious Nature': in either case, for them any such personal attitude cannot have anything to do with the intellectual integrity and success of physics. From the standpoint of the philosophy of wisdom, as we shall see, this division between the personal and the intellectual ought not to be attempted. The intellectual integrity and value of physics itself is intimately associated with its success in expressing and promoting an attitude of love for Nature. Thus Einstein's legitimate intellectual objections to the ultimate acceptability of orthodox quantum theory had everything to do with the high intellectual/personal aspirations that he had for physics – orthodox quantum theory being capable only of claiming 'the interest of shopkeepers and engineers', in that it merely correctly predicts the results of experiments and does not help reveal to us 'the Old One', the architectural grandeur of the universe – orthodox quantum theory thus being, for Einstein, intellectually 'a wretched bungle' (Przibram, 1967, p. 39).

From the standpoint of the philosophy of wisdom, if even something as ostensibly cold and impersonal as physics ought to be pursued as an aspect of love, then most certainly the biological sciences, the social sciences and humanities, and the technological sciences such as engineering and medicine, ought so to be pursued. It is just this which the philosophy of knowledge denies and obscures – and thus, on being put into practice, sabotages. As a result, both the intrinsic intellectual value of inquiry, and the value of our lives, our capacity to love aspects of our world, are undermined.

Quite generally, I wish to argue, our task, in engaging in rational inquiry, is to see, participate in, and help to grow what is significant and of value in existence in the cosmos, questioningly, enjoyably if possible, and above all lovingly.[3] What I mean by this will I hope become clearer as the argument unfolds.

I have now a few remarks to make about the way I expound the argument in the rest of this book.

In chapter 2 I expound the philosophy of knowledge. In chapter 3 I state the basic objection to the doctrine: inquiry pursued in accordance with the philosophy of knowledge fails to satisfy the most elementary requirements for rationality, and as a result must have damaging consequences for almost every aspect of life. In chapter 4 I give a first exposition of the philosophy of wisdom in terms of rational problem-solving. In chapter 5 I expound a somewhat improved version of the philosophy of wisdom formu-

---

[3]In my view it is important to take seriously Popper's point that when it comes to social and political planning, priority should be given to the piecemeal removal of specific cases of avoidable human suffering and injustice, as opposed to the attempt to create an ideal society by means of holistic or Utopian planning. (See Popper, 1961, section 21; and 1969, especially vol. 1, chapter 9.) It is for this reason that I begin with, and lay the greatest stress on, the endeavour to alleviate avoidable human suffering, danger, injustice and death, as opposed to the Utopian endeavour to create 'a more loving world'. However, as we proceed, I shall offer some substantial criticisms of Popper's anti-Utopianism, in chapter 8.

lated in terms of a somewhat improved notion of 'aim-oriented' rationality.

At this point it may be wondered whether it really is the case that it is the blatantly and damagingly irrational philosophy of knowledge, as opposed to the more rigorous and valuable philosophy of wisdom, that at present predominates over the academic enterprise. In chapter 6 I give grounds for holding that the philosophy of knowledge does indeed at present prevail in scientific and academic practice. I then, in chapter 7, assess the basic argument of the book.

In chapter 8 I let the opposition speak. I expound a number of arguments criticizing the philosophy of wisdom and defending the philosophy of knowledge, and I try to rebut these arguments.

In chapter 9 I demolish the philosophy of knowledge in that department of inquiry where it would seem to be the most defensible – physics pursued for its own sake. I argue that the pursuit of knowledge in the physical sciences cannot be dissociated from the pursuit of understanding, from the problematic presupposition that the universe is, in some way, comprehensible. In chapter 10 I tackle the fundamental problem: how can there be life of value granted that the world really is comprehensible more or less as modern science tells us it is?

In chapter 11 I say something about how I myself came to hold the views advocated in this book, and something about the ideas, work and efforts of a number of people and groups during the last decade or so, mainly in Britain, both inside and outside universities, all of which can be interpreted to be a part of a general movement away from the philosophy of knowledge towards the philosophy of wisdom. The revolution that I advocate is already under way!

CHAPTER TWO

The Philosophy of Knowledge

The philosophy of knowledge can be summarized as follows. The proper aim for rational inquiry is to acquire knowledge about the world, objective knowledge of truth. Ultimately, no doubt, knowledge is sought as a means to the end of achieving that which is humanly desirable and of value. At the most fundamental level of all, in other words, the aim of rational inquiry may well be to help promote social progress, human welfare and enlightenment. In order to achieve these fundamental human, social aims, however, it is essential that rational inquiry devotes itself, in the first instance, to achieving the purely *intellectual* aim of acquiring objective knowledge of truth. Only by dissociating itself decisively from the goals, values and beliefs of common social life, so that claims to objective knowledge can be subjected to scrupulously *rational* assessment, can inquiry accumulate genuine knowledge, thus ultimately being of benefit to humanity. Rational inquiry must, as it were, ignore human need in order to help fulfil such need. Truth, not that which is humanly desirable, must be the central intellectual concern of rational inquiry.

Aspects of this basic idea can be traced back to the ancient Greeks, to the Presocratic philosophers, to Plato, Aristotle, Euclid, Archimedes. It is however with the rise of modern science in the sixteenth and seventeenth centuries that the philosophy of knowledge really comes into its own – with the work of Copernicus, Kepler, Galileo, Descartes, Huygens, Hooke, Boyle, Leibniz, and above all with the work of Newton, set out in his *Principia* of 1687. More than anything else, it was the quite unprecedented predictive and explanatory success of Newtonian theory, drawing together and improving on what had gone before, which appeared to demonstrate so conclusively that new, genuine, valuable knowledge about the world can indeed be achieved. A new, assured method for acquiring knowledge had, it seemed, been discovered. The philosophy of knowledge is first and foremost a philosophy of science (here called *standard empiricism*) which, when generalized, becomes a philosophy of all of inquiry. The philosophy of knowledge owes its prestige and influence to being closely associated with the great intellectual success of natural science – in the first instance, Newtonian science.[1]

Francis Bacon, somewhat earlier than Newton, was perhaps the first person to give a clear, powerful and influential expression of the basic ideas of the philosophy of knowledge in something like its modern form. In his writings Bacon stressed the following cardinal points. As a result of acquiring genuine knowledge of Nature, we can enormously enhance our power to act, to do good, to transform the human condition immeasurably for the better. In order to achieve such radical human, social progress, progress in knowledge, in science, is essential. This is to be achieved by means of organized inquiry which bases its results firmly on the ground of observation and experiment, the speculations, prejudices and myths of philosophers and of ordinary social life, of 'common-sense', being firmly ignored (or at the very least not being accepted as true on trust).

There can be no doubt that these ideas, expressed so clearly by Bacon, came to exert a powerful influence over the rise and subsequent development of modern science. Many scientists and thinkers, over the centuries, have been inspired by the Baconian idea that knowledge is of great human, social value. The idea that organized inquiry is needed in order that knowledge may be progressively acquired inspired the founding of the Royal Society, the first official scientific society, having, as it did, royal patronage, and being to some extent a model for subsequent scientific societies. Finally the idea that knowledge is to be acquired by ignoring speculations of philosophers, and instead arriving at results based on observation and experiment, has dominated all subsequent science. The details of Bacon's own methodology for science may be incorrect, and may be ignored by good science, as Popper, for one has stressed. Even Popper, however, a vehement anti-Baconian if ever there was one, nevertheless advocates, as the central tenet of his philosophy of science, a thesis that is central to Bacon's empiricism: *a priori* knowledge about the world being impossible, all scientific claims to knowledge must be assessed solely with respect to experimental

---

[1]For a good account of the deification of Newton in the eighteenth century – associated both with actual scientific research and with popular attitudes towards science – see Gay (1973, vol. 2, ch. 3 and ch. 4, section 2).

success or failure.[2] This Popperian version of Bacon's empiricism is built into the whole structure of modern science, and is accepted as valid by most modern scientists.[3] A version of the idea is to be found in Newton's profoundly influential *Principia*. In his famous rules of reasoning – integral to the plan and argument of the book – Newton formulated his own version of Baconian empiricism.[4]

Another major historical source of the philosophy of knowledge is Descartes' enormously influential dualistic theory of mind and matter. Cartesian dualism divides up reality into two sharply distinct worlds: on the one hand the objective world of fact, matter, physical reality; on the other hand the subjective world of mind, consciousness, personal experience, value. Once this view is accepted (as it was at one time by most scientists in one form or another), it becomes natural to suppose that rational inquiry, science, will reflect in its overall character the sharp split between fact and value, objective reality and subjective feelings and desires, which is asserted to exist in reality by Cartesian dualism. The intellectual standards of the philosophy of knowledge do indeed reflect the Cartesian dualistic view of the world, in this way.

In one respect Descartes was somewhat at odds with the philosophy of knowledge in that he held that reason as well as experience is a source of knowledge. This *a prioristic* methodology was influential for a time, on the Continent at least if not in England. But with the eventual downfall of Cartesian physics and the triumph of Newtonian physics, Descartes' *a prioristic* methodology seemed discredited. *A priorism* lingered on for a time, most notably perhaps, in a modified form, in the thought of Kant. The view had little influence, however, over the development of science. Not only did the position seem to be intellectually indefensible – despite all of Kant's obscure ingenuity. In addition

[2]A basic purpose of Popper's philosophy of science is to defend the central tenet of what I call *standard empiricism* – in turn the central component of the philosophy of knowledge – namely '*the principle of empiricism* which asserts that in science, only observation and experiment may decide upon the *acceptance or rejection* of scientific statements, including laws and theories' (Popper, 1963, p. 54).

[3]That ultimately only observation and experiment can decide the fate of laws and theories in science is a point constantly affirmed by scientists in textbooks, popular lectures and elsewhere, often, as in the case of such figures as Medawar, Bondi, Eccles and others, Popper's insistence on the key role that empirical refutation has for science being enthusiastically endorsed.

[4]'In experimental philosophy we are to look upon propositions inferred by general induction from phenomena as accurate or very nearly true, notwithstanding any contrary hypotheses that may be imagined, till such time as other phenomena occur by which they may either be made more accurate, or liable to exceptions' (Newton, 1962, vol. 2, Book III, Rules of Reasoning in Philosophy, rule IV, p. 400).

the best candidates for *a priori* knowledge – Euclidean geometry and the principles of Newtonian mechanics – were successively dethroned from this exalted position by developments in mathematics and physics: by the development of consistent non-Euclidean geometries in the nineteenth century, and by the development and empirical success of Einstein's special and general theories of relativity in the twentieth century.

By the eighteenth century, Bacon's basic ideas had come to seem, to the thinkers of the Enlightenment, almost commonplace. In essence, Enlightenment thinkers made one vital addition to Bacon's version of the philosophy of knowledge: they stressed the importance of acquiring knowledge of man, of society, of history, in addition to acquiring knowledge of Nature, for achieving social progress, human enlightenment. Thus Vico, Montesquieu, Helvétius, d'Holbach, Voltaire, Diderot, Gibbon and Hume were all concerned, in various ways, to do for man, culture, society, or history, what Newton had done for Nature: to put 'moral philosophy', the study of man, on as sound a footing as Newton had put natural philosophy, the study of Nature: see, for example, Gay (1973).

In the universities in Europe during medieval times, Christianity undoubtedly constituted the dominant philosophy. The basic aims and methods, assumptions and values, of almost all intellectual work was set by Christian doctrine. Throughout the eighteenth and nineteenth centuries, the Bacon-Newton-Enlightenment version of the philosophy of knowledge came more and more to predominate, until by the mid-twentieth century it had come to reign supreme throughout almost all scientific, academic thought and work. The great industrial, technological and medical progress, achieved in the so-called western world at least during the nineteenth and twentieth centuries, intimately associated with scientific progress, seemed to confirm entirely Bacon's vision. Many writers of course continued to stress the importance of other factors for human progress besides progress in knowledge: factors such as faith, morality, imagination, tradition, justice, political liberty, democracy, legal reform, economic progress, industrial development. Some expressed suspicion of the idea that real human progress could be achieved through progress in science and technology. Few however doubted that knowledge is at least necessary, if by no means sufficient, for human progress. No one seems to have challenged the basic tenet of the philosophy of knowledge, namely that rational inquiry should be devoted in the first instance to the achievement of knowledge.

Today the Bacon-Newton-Enlightenment philosophy of know-ledge, suitably qualified, is built into our whole socio-cultural order. It is built into almost all scientific, academic work and thought, and into the way this is related to the rest of society and culture.

I shall now formulate in a little more detail, in the following nineteen points, that version of the philosophy of knowledge, inherited from Bacon, Newton and the Enlightenment, which has come to be embedded, I claim, in the whole intellectual/institu-tional structure of modern scientific, academic work and thought. (What follows, let me emphasize, in order to avoid possible misunderstandings, is my best attempt at a sympathetic exposition of the doctrine that I shall subsequently criticize and reject as irrational, and as intellectually and humanly damaging.)

1   Ideally, the basic aim of inquiry[5] is to produce that which is of human, social value, inquiry thus contributing to human welfare, to human progress, to the quality of human life. In this respect inquiry does not differ from other socially valuable human enterprises, such as theatre, medicine, literature, art, law, industry, education, democratic government. All these enterprises may be held to have the common aim of producing that which is of human value, thus contributing to the quality of human life.

2   The specific aim of inquiry is to produce objective knowledge of truth – and also to provide explanations and understanding. In other words, inquiry contributes to the common human aim of producing that which is of human value by, in the first instance, realizing the distinctively academic or intellectual aim of pro-ducing reliable, objective, factual knowledge, insofar as this can be achieved.[6]

[5]Inquiry is presumed here to be *rational, organized inquiry* – inquiry having something of a public, social or institutional character. We cannot, however, identify inquiry with *science*, since this leaves out of account rational branches of inquiry devoted to the acquisition of knowledge – such as historical research perhaps – which cannot be held to be scientific. Nor can we identify inquiry with *academic* inquiry – since this leaves out of account scientific and technological research, conducted in research institutions or in connection with industry or defence, pursued in accordance with the edicts of the philosophy of knowledge, and yet 'non-academic'. Roughly speaking, the philosophy of knowledge assumes that genuinely rational inquiry is the union of scientific and technological research on the one hand, academic research on the other hand.

[6]In principle a much more modest version of the philosophy of knowledge can be upheld, according to which inquiry eschews altogether the aim of benefiting humanity, the basic aim of inquiry being merely to acquire knowledge irrespective

3   Mathematics, statistics and logic are concerned to improve knowledge of formal, *a priori* or analytic truth. The physical sciences are concerned to improve knowledge about diverse aspects of the physical universe. The biological sciences are concerned to improve knowledge about life. The social sciences and humanities are concerned to improve knowledge about diverse social and cultural aspects of human life. The technological sciences are concerned to improve knowledge needed in order to realize diverse, valuable, practical social goals.

4   In improving our knowledge and understanding of truth, inquiry contributes to the quality of human life in two rather different ways.

First, the contribution is direct. The search for truth is of intrinsic human value, of value when engaged in for its own sake. In pursuing pure research, at either first or second hand, and in observing the scrupulous intellectual standards required in order to pursue such research successfully, we can be spiritually enriched in much the same way in which we can be enriched by taking part in artistic endeavour. Pure science and scholarship, like music, literature and art, contribute directly to our culture, our civilization.

Second, the contribution is indirect. As a result of improving our knowledge and understanding of truth, we may discover how to apply our new knowledge to help realize important human, social objectives, help solve human, social problems. Pure science, in other words, leads to applied science, to technology, which, we may hope, is used in humanly beneficial ways, to help promote human welfare.

Thus, inquiry is of intrinsic or cultural value, when pursued for its own sake, and of pragmatic or technological value, when pursued as a means to the realization of non-academic, human, social ends. To say this is not to say that it is always clear of any particular piece of research whether it is of value culturally or technologically, or both together. Nor should it be assumed that pure research always comes before technological research, technology 'applying' the results of previous pure research. Scientific research that is predominantly technological in character can produce successful results before theoretical explanation and understanding of these

---

of whether this is of value to people or not. It is, however, difficult to justify the modern academic enterprise in terms of this excessively modest version of the philosophy of knowledge. Why should vast sums of public money be spent on organized inquiry if this in no way aims to be of benefit to people? In any case, this modest version of the philosophy of knowledge will be refuted in chapter 9.

results have been achieved. Technological research may even, on occasions, throw up results which lead directly to important theoretical developments, to progress within pure science. It is still possible, nevertheless, to make a sharp distinction between the two ways in which inquiry can be of value: of value in itself, or of value as a means to the realization of non-academic, social objectives of value, such as health, comfort, communication, transport, etc.

5   The fundamental methodological prescription of the philosophy of knowledge can be formulated like this. It is absolutely essential that the intellectual domain of inquiry be sharply separated from, and preserved from being influenced by, all kinds of psychological, sociological, economic, political, moral and ideological factors and pressures which tend to influence thought in life, in society. Feelings, desires, human social interests and aspirations, political objectives, values, economic forces, public opinion, religious views, ideological views, moral considerations, must not be allowed, in any way, to influence scientific or academic thought within the intellectual domain. Only questions of fact, truth, logic, evidence, experimental and observational reliability and success must be considered. Only those factors must be considered, and allowed to be influential, which are relevant to the determination of truth and the acquisition of knowledge. All additional extra-academic human, social considerations and factors must be ruthlessly held at bay and ignored.[7]

The reason for all this is simple. The fundamental intellectual aim of inquiry is to improve our knowledge of objective, factual truth. We can only hope to achieve this aim if we allow only issues of fact and truth to influence our choice of results and theories.

[7]Popper puts the point like this: 'It is clearly impossible to eliminate . . . extra-scientific interests and to prevent them from influencing the course of scientific research. And it is just as impossible to eliminate them from research in the natural sciences – for example from research in physics – as from research in the social sciences. What is possible and what is important and what lends science its special character is not the elimination of extra-scientific interests but rather the differentiation between the interests which do not belong to the search for truth and the purely scientific interest in truth . . . In other words, there exist *purely* scientific values and disvalues, and *extra*-scientific values and disvalues. And although it is impossible to separate scientific work from extra-scientific applications and evaluations, it is one of the tasks of scientific criticism and scientific discussion to fight against the confusion of value-spheres and, in particular, to separate extra-scientific evaluations from *questions of truth*' (Popper, 1976, pp. 96–7). The classic statement of the need to exclude values from the social sciences is perhaps the one to be found in Weber (1949).

The moment we allow our human desires and values, our political objectives and ideologies to influence the way in which we accept and reject theories and results within scientific, academic inquiry, knowledge of objective fact must inevitably be subverted or corrupted. The objectivity, intellectual integrity, and rationality of inquiry must be undermined. Objective knowledge of truth will degenerate into prejudice and ideology. Scientific, academic inquiry must lose its entitlement to the claim that it achieves and produces authentic, objective knowledge of truth.

It is not just the intellectual integrity of inquiry that is at issue here: the human value of inquiry is at issue as well. For the human, social value of inquiry resides precisely in its capacity to produce genuine objective knowledge of truth. Almost paradoxically, in short, in pursuing inquiry we must, within the intellectual domain, ruthlessly ignore all questions concerning human values and aspirations precisely so that inquiry may ultimately be of genuine human value and may help us to realise our human aspirations.

The Lysenko episode in Soviet biology provides us with a classic illustration of just how disastrous can be the outcome, in both intellectual and human terms, if these simple points are violated. As a result of the imposition of Lamarckian ideas on Soviet biology, for ideological reasons, through external political pressure, not only was Soviet biology severely retarded from an intellectual standpoint; in addition, all this had disastrous consequences for Soviet agriculture, which in turn had harmful human, social repercussions.[8]

6 An important qualification must now be made to what has just been said. It must of course be conceded that in pursuing scientific, academic research, scientists and academics are in fact motivated by the desire to achieve all kinds of 'extra-academic' goals. All kinds of 'extra-academic' factors and considerations influence the research aims that scientists and academics in practice pursue. Academics may pursue research in order to discover what will be of benefit to humanity. Academics may well be motivated by passionate intellectual curiosity, by the desire to achieve lasting recognition, to win the esteem of colleagues – or by the desire to advance an academic career or earn a living. All kinds of moral, social, cultural, philosophical, ideological, economic and political factors and considerations may influence the aims which

[8]Anyone who has any doubts about the appalling and disastrous consequences of Stalin's support for Lysenko should read Medvedev (1969), (1971).

scientists pursue in their research, the problems that academic research seeks to solve. This influence operates roughly as follows. The research problems that are in fact tackled, the research aims that are pursued, by the scientific academic community, are the outcome of (a) decisions of individual scientists and academics as to which problems they seek to investigate, which research aims they pursue; (b) decisions of various research institutions as to what type of research should be undertaken; (c) decisions of funding bodies as to which scientists, institutions and research projects should receive financial support; (d) more general policy decisions as to what kinds of research should receive financial support. All these types of decisions may be influenced by all kinds of extra-academic personal and social factors and considerations – in addition, of course, to being influenced by scientific, intellectual considerations.

The all important point, however, is that when it comes to the assessment of results, the assessment of potential contributions to knowledge, the assessment of scientific, academic progress, all these extra-academic factors and considerations, aims and desires, must be ruthlessly ignored, adequacy to the facts, the evidence, the truth, alone being taken into account. In the context of research or discovery, all kinds of extra-scientific, extra-academic personal, social, evaluative factors may legitimately influence scientists and academics in their choice of research aims and problems. In the context of justification, verification, corroboration, or assessment of results, however, one aim only must be taken into account, namely the aim to discover truth, to accept that which constitutes authentic, objective knowledge of fact.[9]

[9]A number of historians of science have argued, explicitly or implicitly, that in order to understand the origins and development of modern science it is essential to see science in its personal, social and cultural context, influenced by diverse personal, social, political, economic and religious factors. See, for example, Koestler (1964); Manuel (1968); Merton (1970); Teich and Young (1972); Webster (1975); Mandrou (1978). In addition, sociologists of knowledge and of science have argued for the need to take sociological factors into account in understanding science – either internal to the scientific community or also external. See, for example, Hagstrom (1965); Ben-David (1971); Mulkay (1979). Others have been concerned to point out the important role that various kinds of human interests play in science, and have done so, either tediously and tendentiously, as in the case of Habermas (1972), or brilliantly, with a wealth of factual detail, and with real moral concern, as in the case of Greenberg (1971). Others again, following Francis Bacon, have argued in various ways for the need for science to give greater priority to helping to promote human welfare, justice and liberation: for example, Bernal (1967); Ravetz (1971); Easlea (1973). None of this diverse work, and none of these diverse arguments, in any way goes against the basic claim that science ought to be

All this might be summed up as follows. The intellectual domain of inquiry must be shielded from the potentially corrupting influence of a largely irrational society if inquiry is to retain its rationality, objectivity, intellectual integrity, entitlement to the claim that it produces genuine knowledge. It must, of course, be recognised that inquiry forms a part of the social order, all kinds of social factors influencing the aims of research. This influence is entirely harmless, however, as long as results are assessed entirely with respect to truth and their capacity to contribute to objective knowledge, social influences here being excluded.

This basic requirement, that a sharp demarcation be maintained between the intellectual domain of inquiry and the rest of society, has a number of further consequences, some of which are now indicated.

7   The intellectual *aims* of scientific, academic inquiry must be sharply distinguished from the personal, social aims implicit in much scientific, academic research. The intellectual aim of all scientific, academic inquiry is, quite simply, truth, the attainment of objective knowledge of value-neutral truth, together with the development of theories which successfully predict and explain factual truth. The personal, social aims of scientific, academic inquiry, on the other hand, may be manifold, as has been indicated above.

8   The intellectual *problems* of scientific, academic inquiry – scientific, academic problems – must be sharply distinguished from human, social problems. Intellectual problems arise when we do not know how to achieve the basic aim of inquiry, namely

pursued in accordance with the edicts of the philosophy of knowledge and, indeed, is, and has been, so pursued. The one essential requirement that must be fulfilled, if science is to be pursued in accordance with the philosophy of knowledge, is that results of research are assessed solely with respect to truth and falsity. This clearly leaves endless room for personal, social and cultural factors to influence choice of research problems, aims and priorities – and thus to influence what science comes to develop knowledge about. How much science is done, public attitudes towards science, the use that is made of new knowledge, the very adoption of the philosophy of knowledge itself by the scientific community: all these vital aspects of science in any society constitute social and cultural aspects of science inevitably linked, in one way or another, to other aspects of the given society. In brief, many critics of scientific and academic orthodoxy have at most criticized only extremely crude versions of standard empiricism and the philosophy of knowledge, and have left uncriticized and unexamined more sophisticated versions of these doctrines that exercise such a profound, and damaging, influence over so much scientific and scholarly work.

knowledge of truth. Intellectual problems emerge when we discover that our knowledge is defective or incomplete in some way – when theories and experimental results conflict, or when experimental results receive no satisfactory theoretical explanation. Human, social problems, on the other hand, arise when we do not know how to achieve human, social goals – enough to eat, good living conditions, health, friendships and love, justice, peace, a productive and creative way of life, happiness.

Intellectual problems have an impersonal, objective character, in that they can be conceived of as existing relatively independently of the particular thoughts, experiences, aims and actions of individual people. Human, social problems, on the other hand, are essentially problems experienced and confronted by people in their lives. Such problems cannot be detached, as it were, from the actual thoughts, experiences, aims and actions of people in society.

9   Just as intellectual aims and problems must be sharply distinguished from social aims and problems, so too intellectual *progress* must be sharply distinguished from social progress. Intellectual progress has to do exclusively with the extent to which science, or inquiry more generally, acquires knowledge and understanding of truth, in an entirely impersonal, objective sense. Thus the intellectual progress of science is not to be assessed in terms of the extent to which *people* enhance their personal knowledge, understanding and appreciation of the world around them, or the extent to which this enriches their lives. (All this is a question of psychological and sociological change, provoked perhaps by the advancement of science, but not itself to be identified with the intellectual advancement of science.) The intellectual progress of science is quite distinct from any increasing human value of science, from the capacity of science to promote social progress. Intellectual progress is to be assessed solely in terms of the extent to which intellectual aims are being realized, intellectual problems are being solved. Thus it is entirely possible for science itself, on the intellectual level, to be making great strides forward even though the human value of science is decreasing, and the tendency of science to help promote human progress is decreasing. We may well hope and believe that scientific progress helps lead to human progress. Human progress is not, however, a part of the definition of scientific progress.

10   Rationality, intellectual standards, scientific, intellectual

criteria of acceptability – as these arise within the context of inquiry – are concerned exclusively with the assessment and evaluation of claims to knowledge, the assessment of results with respect to truth, adequacy to fact. Rationality, intellectual standards are in no way concerned with the assessment and evaluation of the personal, social, moral dimensions of the aims of research. Scientists and academics may well hope that good human, social, moral aims motivate scientific, academic research actually being pursued, and that good moral, social considerations influence where relevant the choice of research aims in appropriate ways. The evaluation and assessment of the human, social, moral dimension of the aims of scientific, academic research lies, however, beyond the scope of scientific, intellectual standards, beyond the scope of reason.

Quite generally, in fact, the evaluation and assessment of personal feelings, desires, aims and moral views lies beyond the scope of scientific, academic rationality. In particular, wisdom – being intimately associated with the personal and the evaluative – lies beyond the scope of rationality. It is no part of the intellectual aim of intellectual inquiry to enhance our wisdom.

None of this should be taken as implying that personal feelings, desires, moral views – and wisdom – are of no importance for life, and even for science. Quite the contrary. It is just that these things lie beyond the range of scientific, academic rationality, outside the scope of intellectual standards, which are concerned exclusively with the assessment of truth and claims to knowledge.

11 At the centre of the philosophy of knowledge, forming the paradigmatic core of the doctrine, there is a more specific philosophy of *science*, here called *standard empiricism*. All that the philosophy of knowledge asserts about inquiry as a whole, standard empiricism also asserts about science; and in addition it makes the following crucial assertion: when it comes to the assessment of results in science, the assessment of scientific propositions, laws and theories, these results must be accepted and rejected solely with respect to empirical success and failure, to the justice that they do to observational and experimental evidence, in an impartial way. For a time, perhaps, in science, choice of theories may be biased in the direction of some untestable metaphysical conjecture about the world, some paradigm or 'hard core', in the kind of way described by Kuhn (1962) and Lakatos (1970). In the end, however, empirical success or failure alone must decide the fate of scientific theories. In the context of

discovery, of course, scientists may quite legitimately be influenced in their thinking, their choice of problems and conjectures, by all sorts of extra-empirical, non-rational considerations – philosophical or metaphysical ideas, even personal, religious, political or economic interests and considerations (and it is in this way that social and cultural factors influence the development of science). The crucial point is that when it comes to the context of justification or assessment, one consideration alone must ultimately be permitted to determine acceptance or rejection: adequacy to the data of scientific observation and experimentation.

A rationale for adopting this basic tenet of standard empiricism can readily be given. We do not and cannot possess *a priori* knowledge about the world, secure knowledge arrived at independently of all experience. Only by comparing our theories about the world with the world itself via our experience of it in an ultimately wholly unbiased, impartial fashion, can we hope to improve our scientific knowledge about the world. Thus a discipline which *permanently* biases the selection of theories in the direction of some untestable metaphysical conjecture, upheld in an *a prioristic* fashion, and despite the persistent empirical failure of theories selected in this fashion, cannot procure authentic knowledge, and cannot be scientific. Such a discipline can only produce dogma, ideology, or religious faith.

12 Standard empiricism and the philosophy of knowledge both require, quite essentially, that a sharp distinction can be drawn between (a) the context of discovery, and (b) the context of justification (the context of the appraisal of theories or results from the standpoint of truth). If such a sharp distinction can be drawn, then it becomes intelligible at least to assert that the scientific character of science, and the rational character of rational inquiry more generally, are bound up only with the way potential contributions to knowledge are *assessed* with respect to truth, in the context of 'justification'. It becomes intelligible to hold that all sorts of extra-rational, and even irrational, psychological and social factors may entirely legitimately influence scientific and rational thought in the context of discovery, the context of inventing new ideas, choosing research aims and problems, without the rationality, the intellectual rigour, of science or inquiry thereby being undermined. All this becomes highly suspect, more or less unintelligibly, the moment the possibility of drawing a sharp distinction between the contexts of discovery and justification is cast into doubt. Classic statements of

the importance of distinguishing sharply between these two contexts, reason applying only to 'justification' and not to discovery, are the following by Reichenbach and by Popper. '. . . the way, for instance, in which a mathematician publishes a new demonstration, or a physicist his logical reasoning in the foundation of a new theory, would almost correspond to our concept of rational reconstruction; and the well-known difference between the thinker's way of finding this theorem and his way of presenting it before a public may illustrate the difference in question. I shall introduce the terms *context of discovery* and *context of justification* to mark this distinction. Then we have to say that epistemology is only occupied in constructing the context of justification' (Reichenbach, 1961). 'The initial stage, the act of conceiving or inventing a theory, seems to me neither to call for logical analysis nor to be susceptible of it. The question how it happens that a new idea occurs to a man – whether it is a musical theme, a dramatic conflict, or a scientific theory – may be of great interest to empirical psychology; but it is irrelevant to the logical analysis of scientific knowledge. This latter is concerned not with *questions of fact* (Kant's *quid facti?*), but only with questions of *justification or validity* (Kant's *quid juris?*). Its questions are of the following kind. Can a statement be justified? And if so, how? Is is testable? Is it logically dependent on certain other statements? Or does it perhaps contradict them? In order that a statement may be logically examined in this way, it must already have been presented to us. Someone must have formulated it, and submitted it to logical examination. Accordingly I shall distinguish sharply between the process of conceiving a new idea, and the methods and results of examining it logically. As to the task of the logic of knowledge – in contradistinction to the psychology of knowledge – I shall proceed on the assumption that it consists solely in investigating the methods employed in those systematic tests to which every new idea must be subjected if it is to be seriously entertained' (Popper, 1959, p. 31).

13   There is general agreement amongst proponents of the philosophy of knowledge that the empirical sciences can be ordered into a rough kind of hierarchy. At the bottom, at the most fundamental level of all, we have theoretical physics, and closely associated with it, cosmology. Ascending, we have the theoretically less fundamental parts of physics such as solid state physics and physical chemistry; a little higher, we have the whole of inorganic chemistry, and alongside chemistry astronomy, astro-

physics and the earth sciences (all specialized applications of physics and chemistry). Ascending still higher, we have the biological sciences with organic chemistry, molecular biology, biophysics and biochemistry at the base, sciences such as zoology, botany, anatomy, neurology, genetics half way up, and ecology and the study of animal behaviour at the top. Higher still, we have the social sciences, anthropology, sociology, psychology, linguistics, economics, political science and history.

According to one view – reductionism – we should seek to reduce all these sciences, at least in principle, to theoretical physics. According to a rival view – anti-reductionism – this is either an unrealizable goal, an undesirable goal, or both.[10] The important point is that both views agree that the empirical sciences can indeed be hierarchically organized along the lines indicated, with what is intellectually and explanatorily fundamental at the bottom, each science becoming progressively less and less intellectually fundamental as we ascend to the top. What this means is that a science at one level presupposes and, where relevant, uses the results of sciences at lower, intellectually more fundamental levels, whereas the reverse is not the case. Theoretical physics does not presuppose or use theories from sociology, whereas sociology constantly uses, even if only in an obvious and crude way, theories and results of physics (such as the existence and persistence of gravitation). Or to take less extreme examples, chemistry presupposes physics (especially the theory of atomic and molecular structure and quantum theory) whereas fundamental theoretical physics presupposes and borrows nothing from chemistry (apart occasionally from a piece of chemical technology for instruments, which is another matter altogether).

14 Rather more controversially, a somewhat analogous hierarchial ordering can be discerned within the logical and mathematical disciplines. At the base there is logic. A little higher up, there is set theory. Almost the whole of the rest of mathematics can be interpreted as amounting to more or less specialized applications of set theory.

15 Intellectually respectable inquiry is, according to the philosophy of knowledge, almost entirely to be identified with

---

[10]For a clear, radical statement of the reductionist, physicalist position, see Smart (1963). For anti-reductionist views see Koestler and Smithies (1969); Popper and Eccles (1977).

professional, expert, scientific, academic inquiry. It must, of course, be conceded that intellectual inquiry of a kind does go on outside universities and research institutions, in society, as an integral part of our lives. We are all, all the time, improving our knowledge and understanding of various aspects of the world that concern us, as we live. Such personal intellectual inquiry hardly deserves, however, to be esteemed very highly from an intellectual standpoint, just because by and large it does not satisfy the kind of intellectual criteria that have been spelled out above. Our personal thinking is hopelessly intermingled with our personal lives, our actions, desires, feelings, prejudices, values. Such personal thinking is to be construed as a legitimate object of study for academic psychological and sociological research, rather than itself being an important part of intellectually respectable scientific, academic inquiry.

16 Professional, expert scientific, academic inquiry is thus, according to the philosophy of knowledge, in a position to deliver authoritative judgements concerning questions of fact and truth – where knowledge has indeed been established. People in society who are not experts, not scientifically, academically qualified, not themselves engaged in scientific, academic research, cannot be expected to provide cogent, authoritative criticisms of scientific, academic results, arising out of their own personal views. Academics cannot be expected to treat those who are not academically qualified as *colleagues* propounding ideas, theories, arguments, criticisms that need to be taken seriously, on the intellectual level. The ideas of people in society can of course be *studied* empirically, by psychology and sociology: such ideas do not themselves however, constitute serious contributions to scientific, academic knowledge.

Scientists and academics are only entitled to their special scientific, academic authoritativeness insofar as they restrict themselves to delivering purely factual judgements, judgements concerning truth, that lie within their own particular speciality, their field of academic competence. The moment academics deliver themselves of value judgements, moral or political judgements, they cease to speak in a scientifically, academically authoritative fashion, and speak simply as human beings, as citizens.

17 Insofar as we seek to conduct our own personal thinking in an intellectually respectable fashion, we must endeavour, according

to the philosophy of knowledge, to make our thinking conform, on the personal level to the general principles of intellectually acceptable scientific, academic thought, that have been spelled out above. We must seek to set up a sharp distinction, within ourselves, between our thinking and reasoning, on the one hand, and our emotions and desires, on the other hand. Our beliefs, our knowledge, that which we accept as true, must be subjected to the same kind of impartial, objective, intellectual appraisal found within scientific, academic inquiry at its best, all considerations of personal feelings, desires, aims and values being ruthlessly ignored. The mind must be sharply separated off from the heart, promptings of the heart not being allowed to influence what is accepted by the mind. A main purpose of education is to encourage students to acquire the capacity to appraise ideas in this kind of impersonal, objective, rational fashion.

18   Literature, and art more generally, according to the philosophy of knowledge, make no kind of direct contribution to the intellectual domain of intellectual inquiry. Great literature and art may perhaps have some kind of inspirational value for some brands of intellectual inquiry: they do not, however, have any kind of direct rational contribution to make to intellectual inquiry just because literature and art do not contribute to *knowledge*. In literature, ideas, feelings, values and imaginary human actions are almost invariably interspersed with one another in a complex fashion, as in life. In addition, our emotional responses to literature have a great deal to do with our assessment of its cultural value. Literature does not seek to improve our knowledge of truth, and does not seek to comply with basic intellectual requirements of a search for truth. Literature and art may, however, of course, themselves be legitimate objects for intellectual inquiry, about which we may seek to develop factual knowledge.

19   Ideas, in order to be capable of objective rational appraisal, must be entirely factual in character, capable of being true or false, and thus potential contributions to knowledge. Thus religious views, ideologies, social and political policies, personal philosophies, which intermingle judgements concerning facts and values in an essential way, are incapable of objective, rational assessment and have no place within the intellectual domain of scientific, academic inquiry (though of course factual theories about religious views, ideologies, etc., in fact held in society, do

have such a place). All such ideas may be said to be, in an important sense, irrational.[11] In particular a 'philosophy' of some enterprise, a view about what ought to be, ideally, the basic aims and methods of the enterprise, has no place within the intellectual domain of scientific, academic inquiry. For such a 'philosophy', being a view about ideal aims and methods, must inevitably intermingle factual and value judgements.

It deserves to be noted that both the philosophy of knowledge and the philosophy of wisdom are philosophies of inquiry in this common-sense conception of 'philosophy', in that both are views about what ought to be, ideally, the basic aims and methods of inquiry. Thus neither of these 'philosophies' can have any very respectable place within the intellectual domain of inquiry pursued in accordance with the philosophy of knowledge. In this way the philosophy of knowledge preserves itself from criticism. Once the philosophy of knowledge is adopted and put into practice, accepted intellectual standards effectively debar critical, rational discussion of philosophies of inquiry. Claims to knowledge can be critically discussed: but the adoption of the aim to acquire knowledge as the basic aim for inquiry becomes more or less immune from critical reconsideration.

In an analogous, somewhat more limited way, standard empiricism, once adopted and put into practice by science, preserves itself from effective criticism within science. For standard empiricism implies that only testable factual hypotheses deserve consideration within science. Standard empiricism is not itself a testable factual hypothesis: hence it ought not itself to be critically discussed within science. Discussion of rival philosophies of science must be sharply separated off from science itself, if science is to retain its intellectual integrity as science. And indeed such discussion is at present by and large confined to the ineffective intellectual ghetto of 'the philosophy of science', where it has little hope of influencing aims and methods actually adopted in scientific research.[12]

---

[11]This point is well made, in connection with ideology in Harris (1968, ch. 1).

[12]One contemporary scientist, no doubt expressing feelings shared by many of his fellow scientists, puts the matter like this: ' . . . "the Philosophy of Science" nowadays . . . [is] arid and repulsive. To read the latest symposium volume on this topic is to be reminded of the Talmud, or of the theological disputes of Byzantium. It is not now a field where the amateur philosopher may gently wander and pick a few nosegays. It is fiercely professional and technical and almost meaningless to the ordinary working scientist . . . This is doubly unfortunate: the divorce of Science from Philosophy impoverishes both disciplines' (Ziman, 1968, p. 31).

It should perhaps be noted, finally, that proponents of the philosophy of knowledge can quite consistently acknowledge that 'reason' can be applied to actions and decisions quite generally, and does not have to be restricted in its application to its basic (philosophy-of-knowledge) task of assessing claims to knowledge. There is in fact an extensive literature on such topics as rational decision theory, practical reasoning, rational action. (See Morgan-stern and von Neumann, 1944; Jeffrey, 1965; Borger and Cioffi, 1970; Wilson, 1974; Raz, 1975, 1978; Harrison, R., 1979.) Two requirements must however be satisfied if this is to conform to the philosophy of knowledge. First, 'rational' decision-making and 'rational' action must conform to the edicts of the philosophy of knowledge to the extent that (a) it is based on 'rationally' obtained knowledge and (b) rules of reason – such as the demand for consistency – are themselves of the type stipulated by the philosophy of knowledge (having to do with the assessment of claims to knowledge). Second, research into such topics as decision theory, practical reasoning and rational action must itself conform to the edicts of the philosophy of knowledge: it must seek merely to improve knowledge about these topics. On the whole, the existing literature on these topics does indeed satisfy these two requirements.

What has been spelled out in this chapter is summarized in the following two diagrams. Figure 1 depicts the intellectual, rational-istic aspects of inquiry, as conceived by the philosophy of knowledge, and figure 2 depicts how the intellectual domain of inquiry is conceived to be both dissociated from, and yet influenced by, and influential upon, the social world.

This completes my exposition of the philosophy of knowledge.

My claim is that during the last 400 years or so, with the gradual decline in influence of Christian thought in the universities, something like the above conception of intellectual inquiry has progressively become the predominant creed, so that today it exercises a profound and far-reaching influence over almost every part and aspect of science, scholarship, technological research and education. This conception of what ought to be the basic aims and methods of inquiry has shaped the whole way in which scientific, academic inquiry has developed in the so-called western world, so much so that it is now built into the whole intellectual/institutional structure of the academic enterprise, and the way this is related to life, to the rest of the social world. And it is not just science, scholarship and education that are influenced by the philosophy of

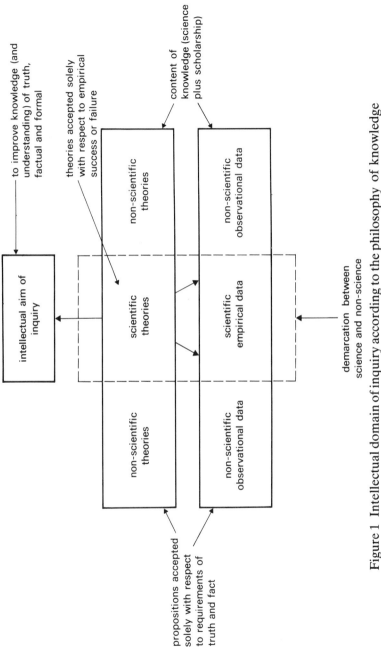

Figure 1 Intellectual domain of inquiry according to the philosophy of knowledge

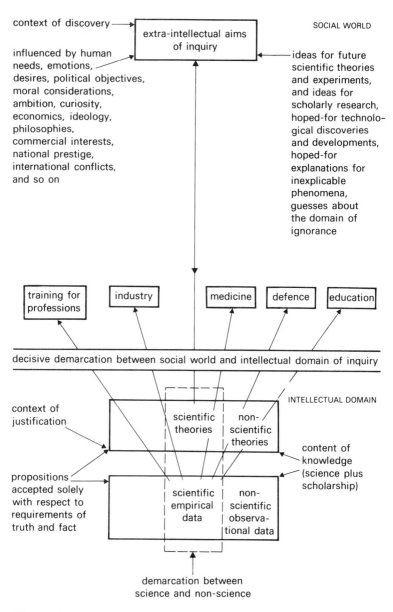

Figure 2 Relationship between the intellectual domain of inquiry
and the social world according to the philosophy of knowledge

knowledge: through these, the philosophy of knowledge exercises its influence, to a greater or lesser extent, over almost every aspect of our personal and social lives. Our very psyches, the way personal thought, feeling, desire and action tend to be inter-related, are affected by the prevalence of the philosophy of knowledge. The central, most urgent point I have to make is how-ever this: quite generally, our overall capacity to realize what is of value in life is adversely affected by the fact that it is the philosophy of knowledge, and not the philosophy of wisdom, that predominates in our world as the official ideal conception of rational inquiry.

In order to appreciate just how massive and extensive is the influence of the philosophy of knowledge over present-day academic work and thought, and over the rest of life, it is essential to consider what academic work and thought would be like if based on the edicts of some radically different philosophy of inquiry. At present, just because the philosophy of knowledge is so widely presumed to be the only sane possiblity, its ubiquitous influence becomes visible. Only in Chapter 6, with the rival philosophy of wisdom before us, will we be in a position to consider seriously the question of which philosophy predominates in practice. For the time being, a few preliminary remarks must suffice.

One strong indication that the philosophy of knowledge does indeed prevail throughout the academic enterprise in practice, is that those who write about the nature of inquiry – philosophers, scientists, sociologists, historians and philosophers of science – almost invariably presuppose in their work some version of the philosophy of knowledge, at least as an ideal of rational inquiry. Furthermore, the philosophy of knowledge is usually presupposed in an unconsciously dogmatic, blind, irrational manner, as if any alternative ideal for rational inquiry is utterly inconceivable.

This claim may well be doubted, in view of the great range of problems tackled, views expressed, and debates engaged in, by those who write about science – or about inquiry more generally. I therefore now set out to show that the philosophy of knowledge, more or less as characterized in the above points, is indeed presupposed – often in an entirely dogmatic way – by a wide range of problems, views and debates to be found expressed in the literature about the nature of inquiry.

Consider first some of the intellectual or philosophical problems that arise once the philosophy of knowledge is presupposed as an ideal of rational inquiry. There are problems concerning the nature of knowledge. Is knowledge to be conceived of in purely intellectual or propositional terms – or in psychologistic, behavioural, social or institutional terms? There are problems about how knowledge is to be distinguished from mere speculation, belief or prejudice, and how science is to be distinguished from non-science and pseudo-science. Do the diverse branches of social inquiry constitute science in the same sense as the physical or biological sciences? Ought precisely the same methods to be employed? There are problems concerning the possibility of sharply distinguishing the contexts of discovery and justification. There are problems concerning the possibility, and even the desirability, of excluding all value-judgements from the intellectual domain of inquiry. There are problems concerning the nature of mathematical knowledge. Does mathematics embody knowledge of a realm of abstract entities independently of human thought? If not – if pure mathematics is not about anything that really exists – how can mathematics be held to embody knowledge at all?

In addition, a number of apparently insoluble problems confront standard empiricism, the paradigmatic core of the philosophy of knowledge. There is the problem of induction. How can scientific knowledge about the world be acquired if theories are selected by means of empirical considerations *alone* (as standard empiricism demands)? There will always be infinitely many different theories which agree equally with all available evidence, but which disagree about unobserved phenomena (at times, places or physical conditions not yet put to the test): what rationale can there conceivably be for selecting just *one* of these theories (by means of empirical considerations alone) as alone embodying knowledge, conjectural or otherwise? There is the problem of verisimilitude. Entirely irrespective of whether we can *know* that science makes progress, what can it *mean* to assert that science makes progress towards increased knowledge of truth, when all scientific theories (so far) strictly are *false*, and no sense can be made of the idea that one false theory is closer to the truth than any other? There are problems of simplicity in science. It is generally agreed that scientists quite properly choose simple rather than complex theories, other things being equal. But what is simplicity here? Is not the simplicity or complexity of a theory something aesthetic and subjective, at best entirely dependent on the language that is used to formulate the theory? If judgements concerning the simplicity or complexity of theories are subjective or language-dependent in this way, how can such judgements have a legitimate role to play in science in influencing which theories are to be accepted and which rejected? Again, if scientists do persistently prefer simple to complex theories, does not this mean that they are in effect prejudging the universe itself to be in some sense simple, permanently biasing choice of theory in the direction of the metaphysical doctrine that nature is simple, thus violating standard empiricism and turning science into a sort of religious dogma? How can permanent preference for simple theories be reconciled with the claim that theories are selected *impartially* with respect to empirical success or failure? There is the problem of the miracle of scientific progress (see for example, Wigner (1970, pp. 222–37). In dreaming up new theories scientists have, it seems, potentially infinitely many embryonic ideas to choose from, to try to develop into fully fledged theories to be put to the test of observation and experiment. The chances of hitting upon an idea that subsequently turns out to be empirically more successful than its predecessors would seem to be infinitely remote. And yet it is just this infinitely improbable act that has been performed again and again by creative scientists – by people like Kepler, Galileo, Newton, Lavoisier, Dalton, Faraday, Maxwell, Darwin, Planck, Einstein, Bohr, Heisenberg, Schrödinger, Dirac, Watson and Crick, Gell-Mann and Ne'eman, Salem and Weinberg. Often, indeed, the key ideas for successful new theories in physics are invented by mathematicians, apparently while uninterested in, and even in ignorance of, the relevant problems of physics. Thus Apollonius developed the theory of conic sections some 1,800 years before Galileo and Kepler discovered that stones and planets move in conic sections. Gauss and Riemann developed Riemannian geometry many decades before Einstein discovered that gravitation is a manifestation of the Riemannian structure of space-time. Hilbert developed his theory of Hilbert space without any idea that just this is what is needed in order to formulate quantum theory, a discovery made subsequently by von Neumann. It almost seems as if there is a mysterious concordance between the nature of the physical universe on the one hand, and the nature of the human mind, pure human thought, on the other hand. All this is difficult, if not impossible, to understand given standard empiricism. Indeed the central tenet of standard empiricism – that all our knowledge about the world is acquired through impartial testing of theories, there being no rational method of discovery in science – seems all but refuted. Finally there are problems about scientific practice. Successful science just does not seem to be pursued straightforwardly in accordance with the edicts of standard empiricism. All too often, scientists fiercely defend theories far less successful empirically than their rivals, even on occasions theories that are ostensibly decisively refuted, and even sometimes theories that are inconsistent, and which thus cannot conceivably be true: and subsequent developments show that this highly anti-standard empiricist behaviour was indeed scientifically fruitful and correct. All this can be found in the work of Bohr in developing early quantum theory, and in the work of Einstein, in his cavalier dismissal of Kaufmann's apparently decisive refutation of special relativity in 1906, and in his espousal of the inconsistent photon theory of light from 1905 onwards. (See, for example, Jammer, 1966; Holton, 1973; Pais, 1980; 1982.)

The striking point to note is that these problems, which arise once standard empiricism and the philosophy of knowledge are presupposed, are just the central problems of academic philosophy of science, and academic philosophy of inquiry more generally. See, for example, Popper (1959; 1963); Nagel (1961); Hempel (1965); Kuhn (1962); Benacerraf and Putnam (1964); Lakatos and Musgrave (1970); Chalmers (1976); Brown *et al.* (1981); Newton-Smith (1982). This in itself strikingly confirms the point that the philosophy of knowledge is widely taken for granted. (As we shall see, quite different problems arise once the philosophy of wisdom is adopted.)

It is true that most scientists tend to dismiss problems such as these concerning the nature of science and of rational inquiry as unimportant, unworthy of serious attention. But this attitude is itself to be explained by the fact that scientists take standard empiricism for granted. Problems such as the above are non-empirical, unscientific, philosophical: standard empiricism thus decrees that they are to be excluded from the intellectual domain of science! It is in this way that standard empiricism, once accepted and built into scientific institutional practice, effectively protects itself from serious criticism.

The innocent might well conclude from the above list of problems – and especially from the apparently insoluble problems confronting standard empiricism – that no doubt can remain: standard empiricism and the philosophy of knowledge are un-tenable, and an altogether different philosophy of rational inquiry must be developed. (This is of course the view of this book.) Academic philosophy of science is based on exactly the opposite position. The vast body of work done in the field during the last few decades almost unthinkingly takes for granted that acceptable solutions to the above problems must presuppose the overall framework of standard empiricism, and certainly the basic tenets of the philosophy of knowledge. Nothing could illustrate more strikingly the extraordinarily dogmatic, irrational manner in which the philosophy of knowledge is upheld. Most academic philosophy of science, indeed, has served to obscure the fact that standard empiricism, our whole conception of science and of rational inquiry, is in deep intellectual trouble. Thus attention has been focused onto ever more elaborate and technical contributions to ever smaller fragments of the problem of induction – taken to be the problem of justifying the rationality of science in standard empiricist terms. This is true, for example, of most of the 1,130 publications on induction referred to by Kyburg (1970), which appeared mainly in the years 1950–70. Again, attention has been focused onto technical problems of simplicity, the unthinking presupposition being that proposed solutions to the problems can only be acceptable if compatible with standard empiricism. See, for example, Popper (1959, ch. VII, 1963, p. 241); Rudner (1961); Bunge (1961); Ackermann (1961); Barker (1961); Goodman (1972 ch. VII); Davies (1973, chs. 4 and 5); Sober (1975); Hesse (1974 ch. 10). In these ways, attention has been deflected away from the real intellectual and human problems that confront science, and organized inquiry more generally. Above all, attention has been deflected away from the basic problem of this book, indicated in Chapter 1 – the basic problem of the philosophy of inquiry: *what ought to be the overall intellectual aims and methods of inquiry if it is to give us the best possible rational help with realizing what is of value to us in life*? If it is essential to reason to articulate basic problems and propose and criticize possible solutions – and basic to irrationalism to block the doing of this – then most contemporary philosophy of science must be judged to betray reason and embody irrationalism.

In defending scientific orthodoxy, philosophers of science have in effect put into practice what Snow once brilliantly called 'the technique of the intricate defensive' (Snow, 1964, p. 67). Discussion of the problems confronting standard empiricism – the central component of the philosophy of knowledge – becomes so elaborate, technical and abstruse, that the simple and decisive objections to the position are lost sight of by everybody, and the position is preserved by default.

Isaiah Berlin once argued, eloquently and persuasively, that the task of philosophy is to call into question basic presuppositions that dominate both thought and life, usually in unnoticed ways (Berlin, 1980, ch. 1; see also Burtt, 1965). It is just this Berlin conception of philosophy that this book attempts to put into practice. Much traditional philosophy of science must be judged to have done the exact opposite of what Berlin advocates, in that it has not developed valid criticisms of standard empiricism and the philosophy of knowledge, and looked for better conceptions of rational inquiry but, on the contrary, has obscured the need to make such criticisms and innovations, and has made the task of seeing what is wrong and what needs to be done all the more difficult. Scientists, historians, philosophers and sociologists of science have all been too quick to identify rational inquiry with science, and the success of science with the adoption of some version of standard empiricism, so that an attack on standard empiricism is interpreted as an attack on science itself, and reason itself! Even Feyerabend, the licensed court jester of orthodoxy, in effect also makes these elementary mistakes, in that his challenge to orthodoxy takes the predictable form of romantic irrationalism or, as he calls it, methodological anarchism. If standard empiricism must be rejected, Feyerabend in effect presumes along with his opponents, then reason itself must be rejected: see Feyerabend (1975).

Insofar as it has been recognized that problems such as the above do call into question what the basic aim and methods of science ought to be, the response has been, over the years, to develop a number of different versions of standard empiricism – thus further obscuring that it is standard empiricism as such that is the source of the trouble. (In what follows, the terms for the divers positions are in part my own.) There are first of all pre-standard versions of empiricism: (1) *infallible heuristic empiricism*, which asserts that from empirical data alone infallible theoretical knowledge can be arrived at by means of inductive methods; (2) *fallible heuristic empiricism*, which asserts that sound knowledge can be arrived at by induction from empirical data – even though this knowledge may be fallible, and may need subsequently to be revised (Bacon, Newton); (3) *a prioristic empiricism*, which asserts that basic metaphysical principles proved by reason together with empirical investigation suffice to enable us to procure almost infallible scientific knowledge (Descartes, Huygens, Leibniz). Versions of standard empiricism proper begin with acceptance of the thesis that the scientific character of science lies in the way results are assessed, and not especially in the way results are first discovered. Diverse versions of standard empiricism are: (4) *infallible inductivism*, which asserts that once laws and theories have been formulated, they can be securely established as true by being derived inductively from empirical data; (5) *fallible or probabilistic inductivism*, which asserts that inductive verification of laws and theories remains fallible or probabilistic, open to revision: see Herschel (1831); Mill (1843); Jevons (1924); Reichenbach (1938); Hempel (1965); (6) *hypothetico-deductivism*, which asserts that hypothetical laws and theories are to be assessed by means of the empirical verification and falsification of propositions deduced from them, there being no such thing as inductive rules of reasoning from data to theories (Peirce, 1931–58; Schiller, 1917, 1921); (7) *falsificationism*, which asserts, in qualification of hypothetico-deductivism, that there is nothing approaching even tentative verification in science, all scientific knowledge being irredeemably conjectural in character, progress being made only through the empirical falsification of theories (Popper, 1959); (8) *standard theoretical pluralism*, which asserts, in addition to falsificationism, that existing theories can only be severely tested if many rival testable theories are persistently developed, since every genuine test is invariably a crucial experiment attempting to decide between rival hypotheses (Feyerabend, 1965); (9) *paradigmism*, which asserts that initially empirically successful or progressive theories – paradigms or hard cores – are accepted and developed within a research tradition, until a rival paradigm or hard core supporting a rival research tradition becomes more empirically progressive, in which case the new paradigm is adopted (Kuhn, 1962; Lakatos, 1970); (10) *standard implicit craftism*, which asserts that the empirical assessment of scientific results is a craft which cannot be adequately encapsulated in any neat set of explicit rules or methods (Polanyi, 1958; Ravetz, 1971); (11) *standard instrumentalism*, which asserts that knowledge in science is confined to empirical laws and data, it being impermissible to interpret high-level theories as embodying knowledge of an unobservable real physical world – theories being no more than devices for systematizing empirical knowledge, or being implicit definitions of key scientific terms (Duhem, 1962; Poincaré, 1952); (12) *standard theoretical realism*, which asserts that it is legitimate to interpret unrefuted scientific theories as tentative conjectures at least about the nature of unobservable physical reality (Popper, 1963; Smart, 1963).

It must, of course, be acknowledged that this list does not even begin to do justice to the variety and misguided sophistication of standard empiricist thought published in recent years. It must also be acknowledged that a few thinkers have rejected standard empiricism (though not necessarily the looser, broader philosophy of knowledge). Notably Russell (1948), in attempting to solve the problem of induction, argued that science must be interpreted as presupposing that nature is uniform and lawful – a metaphysical doctrine to be presupposed to be true independently of all empirical considerations. Even more notably, Einstein (1973), in his later years, repeatedly affirmed his conviction that the universe is comprehensible, and his conviction that science could not sensibly proceed without presupposing the universe to be comprehensible. These views are closer to the philosophy of natural science to be outlined in chapters 5 and 9 as a part of the philosophy of wisdom.

One standard contentious issue within the philosophy of knowledge is the question of how broadly or narrowly the scope of science is to be conceived. There are many natural scientists who hold that the term 'science' should be restricted to the physical and biological sciences, humanistic disciplines such as psychology, sociology, history or anthropology being unworthy of being deemed to be a part of science in that they have failed to develop sufficiently powerful predictive and explanatory theories.[13] There are many scholars in the humanities who enthusiastically agree with this general view as to where the demarcation line between science and non-science is to be drawn – even if perhaps for somewhat different motives. Historians of ideas, art critics, literary critics, historians in general and others of this persuasion deplore the attempt to turn humanistic studies into empirical science: for them, detailed, sympathetic, illuminating and insightful knowledge and understanding of aspects of human life and human creations cannot, in the nature of things, be achieved by the simplistic, factual, empirical approach of science. The attempt leads only to dull and useless collections of facts or, even worse, to empty theoretical verbiage, that is neither science nor scholarship.[14]

---

[13]Ulam expresses an attitude common amongst natural scientists and mathematicians when he writes: 'In social science, a layman like myself feels that there is no theory or deeper knowledge at the present time. Perhaps this is due to my ignorance but I often have the feeling that by just observing the scene or reading, say, the *New York Times*, one can have as much foresight or knowledge in economics as the great experts. I don't think that for the present they have the slightest idea what causes the major economic or socio-political phenomena except for the trivialities everyone should know' (Ulam, 1976, p. 301). Ziman comes to much the same conclusion, even though he puts it more cautiously and politely, see Ziman (1968, pp. 26–9, 1978, ch. 7).

[14]A good example of someone who holds this sort of view in Isaiah Berlin. See, for

Ranging against this fairly orthodox position, with advocates in both the natural sciences and the humanities, there is the view that scientific method can be, and ought to be, fruitfully employed in such human disciplines as economics, psychology, anthropology, sociology, politics and history. This general viewpoint in turn splits into two opposing camps. On the one hand there are those who, like Eysenk (1965), Skinner (1973), Broadbent (1973), believe that scientific method is the same wherever it is to be employed, the methods of the social sciences thus being the same as those of the physical and biological sciences – impartial appraisal of claims to knowledge by means of empirical data. On the other hand there are those who, like Giddens (1976), hold that the social sciences must adopt methods that are in important ways different from those of the natural sciences. In studying aspects of the human world – as one main argument for this position holds – we study ourselves, that which we in part create, something not encountered within the natural sciences: this important difference between the natural and social sciences requires that the two sorts of science adopt methods that are in important respects different.[15]

In addition there are somewhat more philosophical demarcation debates about what precisely is to be demarcated from what, and for what reason.[16] The point I wish to stress here is that these debates about how science is to be demarcated from non-science are all debates *within* the overall framework of the philosophy of knowledge, unthinkingly taking this framework for granted. Even

---

example, Berlin (1980, 1979). The attitude pervades much of Berlin's work, and is especially marked in his defence and celebration of the significance of Vico: see Berlin (1976).

[15]R. D. Laing puts the matter like this: 'It seems extraordinary that whereas the physical and biological sciences of it-processes have generally won the day against tendencies to personalize the world of things or to read human intentions into the animal world, an authentic science of persons has hardly got started by reason of the inveterate tendency to depersonalize or reify persons' (1965, p. 23).

[16]We may be concerned to demarcate: (a) empirical from non-empirical theories (b) scientific from non-scientific inquiry (c) science at its ideal best from merely competent shading into incompetent science (d) science from pseudo-science (e) knowledge from non-knowledge (f) rational from non-rational or irrational inquiry. We may be concerned not just with problems about how to draw lines of demarcation between these different domains, but in addition with problems of providing a rationale for drawing demarcation lines where proposed. Popper's famous solution to his problem of demarcation suffers from a tendency to collapse together (a) to (d), the traditional problem (e) being treated as if it were no more than an aspect of problem (a) or (b), the whole problem of providing a rationale for the proposed demarcation line being neglected.

those hostile to the spread of scientific method into humanistic disciplines nevertheless take for granted basic tenets of the philosophy of knowledge (and are thus already profoundly, if unconsciously, influenced by a generalization of one conception of scientific method).

Furthermore, a range of different versions of the general doctrine of the philosophy of knowledge have also been developed. Some versions, as propounded for example by Brentano, Frege or Popper, conceive of the intellectual domain of inquiry primarily in an impersonal or non-social way, as an autonomous realm of intellectual entities, such as propositions. Other versions, as propounded, for example, by Locke, Hume or Kant, conceive of inquiry in a rather more individualistic way, in terms of ideas entertained, or judgements made, by individual minds. Then again, other versions, as propounded, for example, by Polanyi, Barnes, Bloor or Ziman, conceive of the intellectual domain primarily in institutional or social terms, as a component or aspect of social life. Some versions of the philosophy of knowledge, as upheld perhaps by Einstein, Popper or Polanyi, stress the fundamental importance of knowledge pursued for its own sake, of so-called 'pure' science and scholarship. Other versions, as advocated by Bernal, for example, stress the importance of the practical, pragmatic or technological aspect of science and knowledge. The philosophy of knowledge is not, as I have noted, committed to the fallacious view, implicit in the terms 'pure' and 'applied' science, that theoretical explanation and understanding of phenomena invariably come before technological knowledge and development. The philosophy of knowledge can accommodate the point that technological knowledge is often developed in the absence of, and before, corresponding theoretical understanding is achieved – technological developments and problems even on occasions directly stimulating subsequent theoretical developments, as in the case of Carnot's contributions to thermodynamics. Some upholders of the philosophy of knowledge may advocate that the search for knowledge be related to definite political ideals and objectives, for example those of socialism, liberalism, or of the free market system (this being permissible from an intellectual standpoint according to the philosphy of knowledge, so it may be argued, as long as only research aims and priorities are at issue). Others may be more concerned to stress that the search for knowledge be kept free from political convictions and programmes. Versions of the philosophy of knowledge may well concede, or even emphasize, that broad social, cultural, political and economic factors play a major role in influencing how science and scholarship develop (in permitting or encouraging the traditions and institutions of 'rigorous' inquiry into matters of fact to develop, in providing financial support for such research, and in influencing choice of research aims and priorities). All such views, however diverse they may be in other respects, nevertheless deserve to be considered versions of the philosophy of knowledge insofar as they hold that the basic intellectual aim of rational inquiry is to acquire knowledge, the intellectual worth of potential contributions to inquiry being assessed solely with respect to the contribution that they make to knowledge of truth, intellectual progress thus being distinct from the capacity of inquiry to promote social progress.

So far in this section I have discussed some intellectual or philosophical problems that confront the doctrines of standard empiricism and the philosophy of knowledge, and some of the orthodox responses that have been made to these problems. But there are also, much more seriously, humanitarian problems – as they may be called – that result from attempting to put these (intellectually defective) doctrines into actual academic practice. The academic enterprise is plagued by a range of cultural, educational, social, political and moral problems that all, in one way or another, have to do with the failure of modern science, technology, scholarship and education to be of value to people in life. There is the problem of the inherent triviality of much science and scholarship, the esoteric, jargon-ridden, specialized character of the research obscuring that it is devoid of any real intellectual or practical value, apart from a tendency to promote careers and flatter vanity. There is the scandal of the priorities of world scientific research, around one quarter of the world's budget for scientific and technological research being devoted to military research, some 95 per cent being spent in, and devoted to the interests of, the developed world. There is, as a result, the tendency of much science and technology to serve the interests of the wealthy and powerful, often at the expense of the interests of the poor and powerless, in this way helping to increase inequality and injustice in the world. There are the horrors that scientific knowledge has made possible – nuclear bombs, intercontinental missiles, the means for chemical and biological warfare. There is the highly suspect role that scientific experts have played in actively promoting the nuclear arms race. There are all our modern problems of depletion of natural resources, pollution, rapid extinction of plants and animals, and the destruction of their natural habitat, caused by population growth and industrial development made possible by science and technology.

Valuable contributions to an improved appreciation of major humanitarian problems associated with modern organized inquiry such as these have been made by Jungk (1960), Barzun (1964), Ellul (1964), Greenberg (1971), Roszak (1970), Ravetz (1971), Higgins (1978), Calder (1981), Zuckerman (1982) and many others. This body of work fails however to repudiate the philosophy of knowledge as a damagingly irrational conception of inquiry which urgently needs to be replaced by a more rational kind of inquiry pursued in accordance with the philosophy of wisdom. Either the discussion is confined to political and moral issues, the framework of the philosophy of knowledge being taken for granted (Greenberg); or some orthodox version of the philosophy of knowledge is propounded (Ravetz); or it is argued that the pursuit of knowledge needs to be committed to socialist as opposed to capitalist goals (Rose and Rose, 1976); or reason is identified with what amounts to the philosophy of knowledge, and it is argued that reason needs to be more severely restricted (Roszak) or even repudiated altogether (Feyerabend) – as if too much reason is the problem, rather than too much of a characteristic kind of irrationality masquarading as reason.

The over-riding impression from all the literature so far discussed, however, is that the two kinds of problems – intellectual problems confronting the philosophy of knowledge and humanitarian problems confronting the actual organized pursuit of knowledge in the world – have little to do with one another. This in itself accords with the philosophy of knowledge edict that intellectual and humanitarian problems must be dissociated from one another.

It is a central tenet of this book that the two kinds of problems are intimately interconnected. As I shall argue in chapter 9, many of the above intellectual problems cannot conceivably be solved within the framework of the philosophy of knowledge. Standard empiricism and the philosophy of knowledge, as ideals of rational inquiry, stand decisively refuted. As the above intellectual problems indicate, it is actually profoundly irrational to try to acquire scientific knowledge about the world by selecting theories solely with respect to empirical success or failure; and more generally, it is profoundly irrational to attempt to help people realize what is of value in life by pursuing knowledge, solving problems of knowledge, in a way which is dissociated from a more fundamental intellectual concern with problems of living. If inquiry is *rationally* to help us realize what is of value to us in our lives, it is essential that it gives intellectual priority to our personal and social problems of living, problems of knowledge and technology being tackled in a way which is intimately associated intellectually with discussion of our problems of living, as the philosophy of wisdom requires. Once inquiry irrationally dissociates problems of knowledge from problems of living, as demanded by the philosophy of knowledge, almost inevitably the pursuit of knowledge will come to suffer from the kind of humanitarian defects indicated above.

The argument of this book, indeed, goes further than this. For my basic argument is that a major root cause of so many of the calamities of the twentieth century that humanity has inflicted on itself – the wars, the concentration camps, the totalitarian regimes, the poverty and starvation amidst plenty, the millions upon millions of lives unnecessarily devastated and destroyed – is our long-standing failure to have developed in the world a genuinely rational kind of inquiry devoted to helping us realize cooperatively what is of most value in life. Our self-inflicted calamities in the end result from our general failure to tackle our common problems in a cooperatively rational way: and this in turn is the consequence of our long-standing failure to develop socially influential traditions of inquiry and education devoted to the promotion of cooperative, rational problem-solving in life. In this way the intellectual disasters of the philosophy of knowledge are a distant echo of the human disasters suffered by so many people. In the circumstances, there can scarcely be any more important task for all those in any way concerned with science, technology, scholarship and education than to help develop a more rational kind of inquiry devoted to the promotion of social wisdom.

Proponents of standard empiricism and the philosophy of knowledge may acknowledge the importance of moral and social problems associated with science: they will not however – and here we see the cunning of the philosophy of knowledge – recognize these problems as in any way calling into doubt the integrity of science itself – or calling into doubt the whole way in which we at present conceive of science, or of rational inquiry more generally. For, of course, these are moral, political and social problems, and as such must be, as the philosophy of knowledge stipulates, entirely dissociated from scientific or intellectual problems of knowledge. As *human being* a scientist may well be concerned about such issues; as *scientist* his task is to concern himself exclusively with problems of fact, truth and knowledge. As one author has recently put it, presupposing the philosophy of knowledge and defending a version of standard empiricism: '. . . it is commonplace to speak of progress, meaning an improvement in the material or the "spiritual" conditions of life. Although that sense of progress is unquestionably important, I shall say virtually nothing about it in this essay. My exclusive preoccupation will be with what I call *"cognitive progress,"* which is nothing more nor less than *progress with respect to the intellectual aspirations of science.* Cognitive progress neither entails, nor is it entailed by, material, social, or spiritual progress' (Laudan, 1977, p.7). The success or failure of science, and of our conceptions of science are, in other words, to be judged solely with respect to the capacity of science to realize intellectual aims, disregarding entirely questions as to whether in pursuing these intellectual aims science helps to increase human happiness, or helps to increase human suffering, unnecessary death and injustice. It is this cavalier dismissal of problems of material and spiritual progress as having nothing to do with problems of intellectual progress – inevitable once the philosophy of knowledge is consistently presupposed – that would have so horrified people like Voltaire and Diderot, for whom science was in essence an engine for the promotion of human enlightenment. Carelessly, unthinkingly, the age of reason, the age of enlightenment, has been betrayed!

I do not claim that everyone associated with the academic enterprise accepts the philosophy of knowledge. Nor do I claim that all scientific/academic work proceeds precisely in accordance with the precepts of the philosophy of knowledge. Actual academic work embodies, perhaps, a mixture of the philosophies of knowledge and wisdom. Furthermore, as I shall argue in chapter 11, during the last decade or so a number of developments have taken place in different disciplines, often unrelated to one another, which taken together can be interpreted as constituting a general movement towards the philosophy of wisdom.

If this last point is correct, then it must also be said that this nascent intellectual revolution, from knowledge to wisdom, has so far proved to be somewhat fragmentary, confused and ineffective. The precise and comprehensive character of the change in intellectual aims and methods that is needed, and the precise reasons why this change is needed, have not so far been very clearly articulated, recognized or understood by those urging change in diverse disciplines towards a more humanitarian or socially committed kind of inquiry. In order to criticize the philosophy of knowledge it is not, for example, sufficient to argue that all knowledge is value-laden. A proponent of the philosophy of knowledge will interpret this to be either a platitude (in that what is being asserted is merely that values influence us in what we decide to acquire knowledge about) or a gross fallacy (in that what is being asserted is that factual knowledge always contains an evaluative component so that the truth or falsity of factual propositions depend on value judgements). Nor, in order to criticize the philosophy of knowledge, is it sufficient to argue merely that science and scholarship ought to be more socially concerned, committed or responsible: proponents of the philosophy of knowledge can readily agree. Indeed it is not sufficient at all merely to criticize the philosophy of knowledge, however cogent and decisive such criticism may be. What is needed rather, is to have in existence a clearly formulated alternative to the philosophy of knowledge that is demonstrably more rigorous intellectually and more useful and valuable socially (at least potentially) than what we have at present. Only when such an alternative has been provided can proponents of the philosophy of knowledge reasonably be expected to abandon their creed. Up till now, as far as I know, such an alternative has not been clearly articulated and defended. And as a result the noble efforts of many individuals in diverse contexts to help develop a more enlightened kind of science, technology, scholarship and education, more intelligently, sensitively and effectively devoted to serving the real interests and aspirations of people in life, remain frustrated, ineffective, misunderstood.

In the circumstances it is not surprising that the philosophy of knowledge (more or less as formulated above) continues to be in practice overwhelmingly the dominant intellectual creed of the academic enterprise, exercising its influence over almost every aspect of science, technology, scholarship and education. It influences such things as: what is to count as a contribution to inquiry; criteria of acceptance of potential contributions for publication in academic journals and books; the kind of criticism that is to be permitted to filter into the intellectual domain of inquiry; the content of academic courses, lectures and seminars; conceptions of scientific and intellectual progress; intellectual values and priorities; the content and style of academic contributions and discussion; the accessibility or non-accessibility of academic discussion to non-academics; the awarding of academic qualifications and prizes; academic careers and promotions; the manner in which intellectual research receives, or fails to receive, financial support; criteria for choice of research aims to be actively

CHAPTER THREE

The Basic Objection to the Philosophy of Knowledge

Some objections to the philosophy of knowledge have already been indicated. I now state what is, in my view, the central, fundamental objection to the philosophy of knowledge. It is, I suggest, both simple and decisive. Inquiry pursued in accordance with the philosophy of knowledge violates the most elementary requirements for rationality conceivable – and as a result inevitably tends, in characteristic ways, to betray the interests of humanity.

An elementary requirement for rationality is that, in seeking to solve problems we (a) articulate, and seek to improve the articulation of, the basic problems we hope to solve, and (b) propose and critically assess possible solutions.[1] To this one might add that when we break up our primary, basic problems into a number of subordinate, secondary problems, we (c) tackle these subordinate problems in close association with our primary problems, so that subordinate and primary problems continue to be relevant to each other as we proceed.

It is just these absolutely elementary, general requirements for rationality which are utterly violated if inquiry is pursued in accordance with the philosophy of knowledge.

For what are the basic problems that inquiry, pursued in accordance with the philosophy of knowledge, is designed to solve? The basic (humanitarian) aim of inquiry, let it be remembered, is to help promote human welfare, help people realize what is of value to them in life – knowledge being pursued as a means to this end. But in order to realize what is of value to us in life, the primary problems that we need to solve are problems of *action* – personal and social problems of action as encountered in

---

[1] "... the method of all *rational discussion* ... is that of stating one's problem clearly and of examining its various proposed solutions *critically*' (Popper, 1956, p. 16).

life. From the standpoint of achieving what is of value in life, problems of knowledge and technology are invariably subordinate and secondary to problems of action. Solutions to problems of knowledge and technology contribute to the realization of value in life by extending our capacity to act.

Thus, if inquiry is to pursue its basic (humanitarian) aim of helping us to realize what is of value in life in a way which accords with the above elementary requirements for rationality, then inquiry must give absolute intellectual priority to the tasks of (a) articulating our problems of action and (b) proposing and critically assessing possible solutions – possible personal and social actions. Furthermore, inquiry must (c) tackle subordinate, secondary problems of knowledge and technology in close association with problems of action, so that problems of knowledge and technology continue to be relevant to those problems of action we need to solve in order to realize what is of value to us in life.

It is just these elementary requirements for rationality that inquiry pursued in accordance with the philosophy of knowledge violates. Far from intellectual priority being given to the tasks of articulating problems of living, proposing and criticizing possible solutions – problems of knowledge and technology being tackled as rationally related subordinate, secondary problems – it is all the other way round: problems of knowledge and technology are tackled in a way that is intellectually dissociated from problems of living, the latter, indeed, being excluded from the intellectual domain of inquiry altogether.

In short, inquiry pursued in accordance with the philosophy of knowledge makes the disastrous intellectual mistake – from the standpoint of contributing to the realization of what is of value in life – of giving sustained attention to subordinate, peripheral problems (of knowledge and technology), while discussion of the primary, problems (of personal and social action) are excluded from the intellectual domain of inquiry altogether.[2]

Inevitably, profoundly undesirable consequences result for all aspects of human life (including inquiry itself) if organized inquiry is irrationally restricted to solving problems of knowledge intellectually dissociated from problems of living – as demanded by the philosophy of knowledge in a misguided attempt to preserve

[2]This objection does not apply to that version of the philosophy of knowledge, referred to in footnote 6 to chapter 2, according to which the basic aim of inquiry is merely to acquire knowledge irrespective of whether this is of human value of not. This modest version of the philosophy of knowledge will however be refuted in chapter 9.

reason. I now indicate six such inevitable undesirable consequences (to be further elaborated during the course of the rest of this book). I also indicate, where relevant, how these consequences manifest themselves *in reality*, as a result of organized inquiry *in reality* conforming to the irrational edicts of the philosophy of knowledge. I have already given some grounds for holding that the academic enterprise does in reality conform to the philosophy of knowledge: further even more substantial grounds will be given in chapter 6 below.

1   *There are profoundly undesirable consequences for the general quality of human life.* If people everywhere are to have their best chances of realizing what is of value to them in life, then it is essential that people everywhere can tackle in cooperatively rational ways their common problems of living. The greater the general failure to do this the more unnecessary human failure, suffering and death there will be in the world. But in order for cooperative rationality to develop as an integral part of living, it is essential that institutions of learning – schools and universities – devote themselves to promoting cooperative rationality in life. It is essential that scientific and technological research, scholarship and education give absolute intellectual priority to the tasks of articulating problems of living, and proposing and criticizing possible solutions. In refraining from doing this – as a result of complying with the philosophy of knowledge – organized inquiry fails to do what it most needs to do if it is to help people everywhere realize what is of value to them in life. In scrupulously restricting themselves in their professional capacity to the pursuit of knowledge, scientists, scholars and teachers ignore their central, most vital professional task: to help promote the co-operative rational search for what is of most value in personal and social life. Simultaneously, reason and humanity are betrayed.

If all questions about what we want, what is of value in life, and what we need to do in order to realize what is of value, were entirely unproblematic and uncontroversial, then it might be reasonable to exclude consideration of such questions from rational inquiry. If such questions somehow lay irredeemably beyond the reach of reason, then it would be necessary to exclude them from rational inquiry. In either of these cases, rational inquiry could only be of benefit to humanity by providing knowledge and technology, as Francis Bacon in effect supposed. But both these suppositions are utterly false. Questions about what problems of living we should try to solve, what actions we should perform, in order to achieve what is of value are profoundly problematic and controversial. They can be, and urgently need to be, tackled rationally, at the very least in accordance with the basic strategies of rational problem solving, (a), (b) and (c) indicated above. It is here, indeed, that our greatest failures lie, and our greatest need for rational learning exists. Almost all our major social problems exist not because of lack of knowledge and technology, but rather because of a general failure to develop in the world traditions of cooperative rational problem-solving and learning devoted to enabling people to realize lives of value and justice. Consider the following major social problems confronting humanity today, already referred to at the beginning of chapter 1: problems of extreme poverty, of disease, malnutrition and starvation experienced by millions of people in the third world; problems posed by the existence of dictatorships maintained by force, all political opposition being suppressed, elementary political and civil rights being annihilated, non-violent critics of the régimes suffering arbitrary arrest, imprisonment, and even torture and execution; problems posed by vast inequalities of wealth and power between people, both within nations, and between nations; problems posed by the spread of armaments, non-nuclear and nuclear; problems of war, within nations and between nations; problems posed by the cold war, the balance of terror, the perpetual threat of the nuclear Armageddon. In order for these problems to be progressively resolved in just and humane ways it is necessary for millions upon millions of people to act in new, appropriate ways, in rational response to the problems, individually and cooperatively. Even when it is necessary to develop new knowledge and technology in order progressively to resolve such problems – knowledge relevant for the assessment of proposals for action, for example, or technology relevant for the curing of disease, for the production of food or for birth control – nevertheless such knowledge and technology only assists the just and humane resolution of such social problems insofar as it helps to make it possible for people to perform appropriate cooperative actions (knowledge and technology on their own resolving no such problems of living). The overwhelming need is for millions upon millions of people to discover how to act in more cooperatively rational ways than at present, in response to their common and differing problems: and it is this desperately important need that organized inquiry ignores when it restricts itself with scrupulous irrationality to the pursuit of knowledge.

It is, of course, true that the existence in the world of a tradition of imaginative, open, public, critical, humane and cooperative discussion of basic human problems of living and how they are to be solved is not in itself *sufficient* to ensure that such problems will be tackled in practice in rational, cooperative, humane ways. Rational discussion does not ensure rational action. The existence of such a tradition is however, I maintain, a *necessary* condition for rational social action. In our vast, complex, diverse, inter-dependent, rapidly changing human world there is no chance that more rational, cooperative, humane ways of tackling our major social problems will develop in the absence of sustained *discussion* of how such problems are to be tackled, diffused throughout the social world. Only by cooperatively imagining and criticizing many possible actions (the heart of reason) can people discover those rare, complex, coordinated actions which permit everyone to benefit. It is just this which makes it a matter of such urgency that organized inquiry should take up its proper, fundamental intel-lectual-social task of helping to promote and sustain such discussion, so that it becomes capable of guiding social action.

At present irrationality in life is everywhere apparent. It is apparent in the lamentable failure of humanity to resolve the appalling social problems just indicated. But quite apart from the persistent horrors of the twentieth century, our general failure to develop cooperative rationality in the world is apparent even in the most democratic and liberal societies in existence today, in that in such societies institutions everywhere are organized on hier-archical rather than cooperative lines, with the few people at the top making decisions that the many are to carry out without question. In Britain government is conducted with absurd secrecy, the population being ruled almost like children. Irrationality is manifest in the crudity of the ideals and creeds that govern people's lives – religious, moral, political, economic – the most influential doctrines often being inherited almost unchanged from the nineteenth or eighteenth century – even further back in the case of religion. Irrationality is even more strikingly apparent in the attitudes that so many people adopt to the creeds that govern their lives. Instead of adopting the rationalist attitude that all such creeds amount to no more than inevitably imperfect proposed solutions to life's problems, to be fiercely criticized and improved on wherever possible, just the opposite attitude prevails. Up-holders of such doctrines – whether religious or political – all too often regard all doubt and criticism as inherently bad and hostile. In science it is a commonplace that progress is achieved because of a persistent endeavour to criticize and improve existing theories.

Everywhere in personal and public life one finds the exact opposite of this: doctrines dominating personal and social life are fiercely protected from criticism and improvement. The result is that we are burdened in our personal and social lives with political and religious doctrines – proposals for living – which have been protected from criticism and improvement since their first advocacy, and which, as a result, are grotesquely irrelevant to our present circumstances and problems. Ideas dominating our lives are treated in ways which violate utterly even the most elementary of requirements for rationality, indicated above. The result is that only pitifully slow, intermittent progress is made in developing ideas more adequate and conducive to the realization of value in life. And the result of this in turn is that our *lives* make only pitifully slow, intermittent progress towards the realization of what is of value.

The academic enterprise bears a heavy burden of responsibility for the persistence of this damaging irrationality that pervades the world in failing, over the last century or so, to develop a kind of organized inquiry wholly devoted to the promotion of rationality in life.

Without doubt this is by far the most serious and general undesirable consequence of pursuing inquiry in accordance with the irrational precepts of the philosophy of knowledge. The remaining five undesirable consequences in effect amount to special cases of the above general undesirable consequence.

2   *There are undesirable consequences for the quality of human life as a result of scientific and technological progress.* In a world where humane cooperative rationality prevails, scientific and technological progress is bound to be beneficial (setting aside unlucky accidents). But in a world where such rationality is largely absent, scientific and technological progress is as likely to lead to human suffering and death as to human good. In a world where there are immense injustices, persistent, violent conflicts between people, and where national and international politics are often conducted at the moral level of gang warfare, the products of scientific and technological progress, however nobly sought, and however potentially beneficial to humanity, will be used to imprison, enslave and kill. Even a mere lack of cooperative rationality in human affairs can have the outcome that new technology, potentially beneficial, is used in ways that cause much unintended and unforeseen human suffering and injustice.

Thus research in pure physics, nobly motivated, led to the

possibility of the atomic bomb. The Manhattan project, motivated originally by the understandable desire to ensure that Hitler should not alone possess the atomic bomb, led to such bombs being dropped on Hiroshima and Nagasaki, and to the subsequent superpower nuclear arms race, which has exacerbated the cold war, and now threatens to destroy the world. The so-called 'green revolution', carried through in order to increase food production so that the hungry may eat, has, in many places, not had this effect owing to economic and social conditions, such as grossly unequal distribution of ownership of land. The development of automation and artificial intelligence, potentially enormously beneficial, nevertheless in practice also threatens to create human suffering in that it helps to create unemployment. Lead is added to petrol to solve the 'knocking' problem of motor car engines, even though ingested lead causes brain damage, especially in young children. Industrialization of the wealthy nations, made possible by science and technology to a considerable extent, makes it possible for the wealthy nations of the world to use up an inordinate share of the world's natural resources, at the expense of the poor in the third world. In this way science and technology, via industrialization, make possible the development of vast inequalities of wealth and power, a state of extreme international injustice.

The scientific community cannot of course be held to be solely responsible for the suffering and death caused by the products of scientific research in the hands of others. It can however be held to be wholly responsible for the fact that scientific and technological inquiry have been developed in such a way that they are dissociated from a more fundamental intellectual concern to promote cooperative rationality in life. This is the great intellectual and moral sin of the scientific community.

3   *There are undesirable consequences for scientific and techno-logical research itself – for the priorities of research.* If there is a general lack of cooperative rationality in the world, not only will this lead to the products of scientific research being used in harmful ways: it will also lead to science itself becoming corrupted, in that the aims, the priorities, of scientific research will come to be corrupted. Instead of the aims and priorities of research being intelligently chosen so as to help relieve human suffering, help promote human welfare, on the contrary, in all likelihood, the aims and priorities of research will come to reflect merely the special interests of the scientific/academic community itself, and the interests of those who have sufficient wealth and power to fund

and guide research. This is almost bound to occur once the search for new knowledge and technology is irrationally dissociated from a more fundamental endeavour to promote cooperative rational problem-solving in life.

Writing in the seventeenth century, Robert Boyle, one of the founding fathers of modern science, had this to say about what he called the 'Invisible College' – a sort of forerunner of the Royal Society, and thus of organized scientific research. 'The 'Invisible College' [consists of] persons that endeavour to put narrow-mindedness out of countenance by the practice of so extensive a charity that it reaches unto everything called man, and nothing less than an universal good-will can content it. And indeed they are so apprehensive of the want of good employment that they take the whole body of mankind for their care.' (Quoted in Werskey, 1978, p. 13.) A modern science and technology that put into practice the spirit of Boyle's Invisible College – thus genuinely devoting itself to the welfare of humanity – would today clearly give priority to the problems and needs of the poorest people on earth. Problems of third-world sanitation, agriculture, malnutrition, disease, housing, transport, education, appropriate technology, would be the central focus of much of the world's scientific and techno-logical research. The social sciences would be centrally concerned with the manifold social, cultural, economic, psychological, political and international problems associated with the plight of the world's poor, especially having to do with the way in which the rest of the world interacts with the third world.

Does modern science succeed in devoting itself to the interests of humanity in this way? The answer must surely be that it does not. As we have noted, something like 95 per cent of the world's expenditure on scientific and technological research supports research conducted in the developed world, being devoted primarily to basic (or 'pure') research, military research, and research related to the economic, industrial and social needs of the developed world. It has been estimated that roughly a quarter of the world's investment in research and development is spent on military research, over half a million scientists working on the development of new weapons. Some 15 per cent of the world's research budget is spent on pure science, much of this being siphoned off into high energy physics, of little conceivable potential relevance or interest to the world's poor. Only a very small fraction of the world's scientific and technological research is devoted to the problems of poor people living in the third world.[3]

[3]For a more detailed discussion of these points see Norman (1981, especially chs. 3 and 4).

It is clear that modern science and technology fail quite lamentably to 'take the whole body of mankind for their care'.

Let me emphasize again that this state of affairs is almost bound to arise once the scientific/academic enterprise suffers from the characteristic kind of irrationality advocated by the philosophy of knowledge. It is of course in any case entirely to be expected that the wealthy and powerful will try to subvert scientific and technological research so that it serves their own interests. However, if the scientific/academic community put into practice the view that its basic intellectual and professional task is to promote cooperative rationality in life, then there would at least be general agreement that it is the professional duty of every scientist and scholar to try to discover and draw attention to the often subversive influence of wealth and power and, where possible, to check and oppose it. From this perspective, scientific and technological problem-solving massively unrelated to problems of living of those whose needs are the greatest, is both irrational and immoral. But once the philosophy of knowledge is accepted and put into scientific/academic practice, exactly the opposite situation prevails. From this perspective, the fact that scientific and technological problem-solving is massively unrelated to, or even in direct opposition to, the most urgent and desperate needs of humanity, does not in any way call into question the rationality or the morality of science itself. For, according to the philosophy of knowledge, the primary – perhaps the only – intellectual and professional obligation of the researcher is to acquire authentic, objective knowledge, unrelated to any programme or ideology for bettering the general condition of humanity. It is indeed a primary duty of the scientist – so proponents of the philosophy of knowledge may argue – to *dissociate* the search for knowledge from any political or ideological programme or viewpoint. This must be done precisely so that science may genuinely serve the interests of humanity – by producing genuine, objective knowledge. Thus as long as modern science and technology produce authentic knowledge and reliable technology, there can be no intellectual or moral failing internal to science itself – even if science and technology happen to benefit the wealthy rather than the poor. To try to commit research to a more humanitarian programme would actually be to *subvert* the objectivity, the intellectual integrity, the scientific character, of science. It would actually go against the real interests of humanity!

A scientist must not fake his results, and any scientist caught doing this will immediately be ostracized by the scientific community. It seem that scientists may, however, with impunity produce all sorts of fake arguments when it comes to gaining funds from research projects, just because, according to the philosophy of knowledge, the whole issue of research aims and priorities lies outside the domain of the rational, the scientifically, objectively discussable and assessable. Leading scientists may employ such intellectually disreputable arguments as that support for research in high energy physics is essential for economic development, essential for the preservation of science and civilization, and far from being ostracized, such scientists, if successful, will be showered with scientific rewards and honours.[4] That which is intellectually and morally disreputable from the standpoint of a kind of inquiry devoted to promoting cooperative rationality in life becomes wholly honourable from the standpoint of the philosophy of knowledge, it being rather systematic criticism of research priorities that becomes intellectually disreputable. In this way, acceptance of the philosophy of knowledge not only blinds the scientific community to the moral and intellectual scandal inherent in the priorities of current scientific research; it has the further effect of transforming legitimate criticism of the *status quo* into a dangerous threat to the objectivity, intellectual integrity, and scientific character, of science. Legitimate criticism is ostracized wholesale as irrational and ideological.[5]

[4]Numerous examples of leading scientists employing such intellectually disreputable arguments as these in order to obtain funds for research are described, with scathing comments, by Greenberg (1971). Greenberg is concerned primarily with the politics and the immorality of pure research in the USA in the years 1945–70. He gives a number of examples of scientists putting forward arguments such as that science 'has now become the basis for the advance of our economy' thereby echoing the litany that proponents of basic research have regularly uttered since the establishment of the science-government partnership at the end of World War II (p. 30). Greenberg goes on to point out how deplorable is the case for supposing that basic research has much to do with economic growth. He also remarks 'the cathedral metaphor occurs repeatedly in the public pronouncements of the statesmen of science, as, for example, in the words of Philip Handler, chairman of the biochemistry department at Duke University, chairman of the National Science Board, and a member of the President's Science Advisory Committee: 'The edifice which is being created by science . . . is fully comparable to the cathedrals of the Middle Ages or the art of the Renaissance . . . ' and Greenberg comments 'that the building of pyramids and cathedrals exacted a monstrous toll from the masses that were supposedly elevated by these edifices is never discussed' (p. 35).

[5]'One of the leading statesmen of pure research privately protested the publication in *Science* of a news article that described the Hindsight report' [a Defence Department report arguing that weapons development gained little from pure research]. 'Description of so heretical a thesis, he felt, was tantamount to advocacy of it' (Greenberg, 1971, p. 31).

A further point is this. Even if there is the desire, and the power, to develop scientific and technological research aims and priorities in directions genuinely of maximum value to humanity, nevertheless it is still highly problematic to make good choices of research aims and priorities. For in order to make such good choices we must bring together good metaphysical, scientific, social and value decisions concerning such things as the domain of our ignorance, what is potentially scientifically discoverable, what is most urgently needed by people, today and in decades to come. We must discover that highly problematic region of overlap between the scientifically discoverable and the humanly desirable. A kind of inquiry that gives intellectual priority to promoting cooperative rationality in life has the capacity to help us make good decisions about these matters, in that it demands that inquiry incorporates explicit, imaginative, and critical discussion of actual and possible aims and priorities for scientific and technological research, rationally related to explicit, sustained, imaginative, and critical discussion of problems confronting people in their lives. The methodology, the intellectual standards, of inquiry of this type are thus designed to help us to develop and choose aims and priorities for scientific and technological research best designed to be of greatest value to people in their lives, to be of value to humanity, with justice.

In contrast to this, the philosophy of knowledge banishes discussion of research aims and priorities – and discussion of life aims and priorities of people – from the intellectual domain of inquiry. In particular, the philosophy of *science* of standard empiricism – the hard core of the philosophy of knowledge – banishes discussion of untestable ideas (metaphysical or evaluative) from the intellectual domain of science. Thus the methodology, the intellectual standards, of the philosophy of knowledge, far from aiding, actually seriously obstruct, the task of discovering and choosing good aims and priorities for scientific research. Instead of there being sustained imaginative and critical discussion of actual and possible research aims as an integral part of scientific discussion – in scientific papers, monographs, textbooks, lectures, seminars and conferences – the debate rather is confined primarily to those with power to decide, such as grant-giving bodies, heads of research institutions, those concerned to determine science policy.[6]

---

[6]Chapter 6 gives an indication of the extent to which discussion of actual and possible scientific research aims, and how these do and might relate to problems and aims of living, is in practice at present excluded from the intellectual domain of science, as defined by 'science abstracts' (see pp. 134–6).

4   *There are undesirable consequences for the cultural (or 'pure') dimension of scholarship and science.* All branches of inquiry in the end owe their *intellectual* value to their capacity to enable people to realize what is of *personal* value in life. This obviously holds for the practical or technological dimension of inquiry: it also holds, I claim, for the cultural dimension of inquiry. Whether the subject is history, anthropology, cosmology, philosophy or pure mathematics, the discipline is of intellectual value, from a cultural standpoint, insofar as it can be used by people to extend their own personal capacity to see, experience, know, understand or appreciate significant aspects of the world, or significant possibilities. Personal (and interpersonal) inquiry is what ultimately matters: impersonal inquiry is but a means to that end. As Einstein once remarked: 'Knowledge exists in two forms – lifeless, stored in books, and alive in the consciousness of men. The second form of existence is after all the essential one; the first, indispensible as it may be, occupies only an inferior position' (Einstein, 1973, p. 80). All inquiry – practical *and* cultural – is in the end to be evaluated from the standpoint of its capacity to enrich human life.

However, if inquiry is to be of value in this way, it is absolutely essential that priority is given, within inquiry, to the *activity* of people (a) articulating their own personal problems of knowledge and understanding, and (b) proposing and criticizing possible solutions. All other more impersonal intellectual problems need to be tackled as elaborations of personal problems, as secondary and subordinate to primary, personal problems of knowledge and understanding. If this is done, then the cultural dimension of inquiry can flourish, in that impersonal intellectual problems of scholarship and science can become sensitively and intelligently responsive to personal problems of knowledge and understanding encountered by people in life. It becomes possible to pursue the humanities and social inquiry in such a way that they are devoted to helping people articulate each other's problems of living and each other's possible and attempted solutions, so that people separated by space, language or culture may enter imaginatively into each other's lives, thus improving mutual understanding – of value in itself, and of value in that it is essential for cooperative rational action. (In the case of history, of course, communication can proceed in one direction only.) In this way, what is potentially best in the humanities and social inquiry can flourish. Equally, it becomes possible to pursue natural philosophy as the cooperative outcome of the passionately personal endeavour to improve personal knowledge and understanding of the natural world.

Children, on being introduced to cosmology, can begin by articulating problems of understanding confronting their own childish cosmological ideas. Proposed solutions to such problems offered by others can also be considered, so that from the outset the ideas of Democritus, Kepler, Galileo, Newton, Darwin, Einstein and others can be used to develop and solve one's own personal problems of understanding. Just as in history or anthropology we endeavour imaginatively to see the world as seen by others, so too in science we can endeavour to see the world as seen by Boscovich, Faraday, Pasteur, Planck or Weinberg, thus improving our own understanding of nature. As a result of using public science in this way, we can come to appreciate both the cooperative, and the passionately personal character of the best of science. We can share in the noble quest to understand revealed in the lives of people like Kepler, Faraday, Darwin or Einstein. The best of what there is in science, from a cultural standpoint, can flourish.

All this is sabotaged when scholarly and scientific research is sharply dissociated from personal problem-solving in life as demanded by the philosophy of knowledge. For, once scholarship and science become dissociated intellectually from the endeavours of people outside universities to improve their personal knowledge and understanding, the vital personal and interpersonal dimension of inquiry tends to disappear from view. Scholarship and science tend to become esoteric, formal, scholastic and decadent, remote from the interests and concerns of non-academic life, pursued for the sake of academic career and status rather than for the sake of shared personal understanding. Social inquiry fails to promote person-to-person understanding between people in the world; natural science fails to promote cooperative, personal understanding of the natural world.

These undesirable consequences of irrationally dissociating scholarly and scientific problem-solving from personal problem-solving in life are, I suggest, everywhere apparent in the modern academic world.

5   *There are particularly undesirable consequences for inquiry, when the object of study is ourselves, aspects of our human world – as a result of social inquiry being pursued as social 'science'.* Instead of helping us to see, to discover, what is of most value in people, in institutions, in artefacts created by people, rather the social 'sciences' (pursued in accordance with the philosophy of knowledge), eschewing 'value' in order to be 'objective', 'factual'

and 'scientific', must inevitably invite us to see people, society and culture, in a value-denuded way, thus obstructing our capacity to see value in life. Instead of giving priority to the task of articulating problems of living, proposing and criticizing possible solutions, the social 'sciences' must rather confine themselves to acquiring knowledge, profoundly influential assumptions about human problems and their possible resolution being placed beyond critical discussion. Not only does this deflect attention from the central task of articulating our problems of living, proposing and criticizing possible solutions. In addition it must tend to lead social 'scientists' to advocate, in a wholly surreptitious fashion, ways of conceiving of our problems and how they are to be solved when they ostensibly only advocate neutral items of social knowledge. For inevitably, built into supposedly purely factual descriptions and theoretical explanations of social phenomena – in economics, sociology, social psychology, educational psychology and so on – there must be some implicit presuppositions about what our problems are, what it is possible and desirable to do in order to solve them, in these fields. Such presuppositions about what really matters remain, however, hidden and repudiated. Thus in seeking to make the social sciences rigorously factual and scientific, social scientists actually help to sabotage rationality in life. In ostensibly propounding factual knowledge they surreptitiously advocate approaches to life and its problems, at the same time in all honesty denying any such thing is being done, the intimated views as to what our problems are and how they are to be solved thus being placed beyond criticism, even beyond recognition. Yet again, instead of seeking to help improve the understanding that people have of each other in society, the social 'sciences' (pursued in accordance with the philosophy of knowledge) must rather give priority to the improvement of knowledge and understanding of people and social phenomena within the social 'sciences' themselves, of a specialized, professional kind, dissociated from and often unintelligible to people in society.

The idea that the chief aim of social inquiry ought to be to develop specialized knowledge of social phenomena may indeed be held to be one of the most seriously damaging implications of the philosophy of knowledge. For it is above all widespread acceptance of this implication which effectively puts a stop to academic inquiry being pursued fundamentally as the promotion of humane, cooperative problem-solving in life.

The situation is, however, in some respects even worse than this might suggest. Not only does general acceptance of the philosophy of knowledge prevent social 'scientists' from taking as their fundamental task the promotion of rational cooperative human problem-solving in society: even worse, it leads social 'scientists' to produce work which, if anything, serves actually to encourage social manipulation – thus further obstructing rational, co-operative, social action.

As far as the natural and biological sciences are concerned, it is quite clear that improved theoretical knowledge of natural phenomena does on occasions lead to valuable new technology. This depends crucially on improving our knowledge of the *lawfulness* of natural phenomena. Our knowledge predicts that if such and such conditions obtain, such and such will reliably be the outcome. Knowledge of this type enables us successfully and reliably to manipulate natural phenomena to our advantage.

The philosophy of knowledge in effect takes for granted that a similar procedure ought to be attempted as far as the social sciences are concerned. The human, moral, implications of this, however, are horrifying. For in essence what is being advocated is this. First, the social sciences need to develop improved theoretical knowledge of laws governing human behaviour and social systems. This knowledge then enables us to predict that if such and such human, social circumstances are realized, such and such will reliably be the outcome. As a result, we are in a position to develop useful social technology. But this amounts quite simply to developing techniques of human, social manipulation. Built into the very enterprise of the social sciences, conceived of in this way, is the ideal of developing more effective techniques for mani-pulating people.[7]

The saving grace of this procedure is perhaps its ineffectiveness. People are not just natural phenomena. Human actions are not law-governed in the way in which natural phenomena are (even though all physical processes occurring in connection with human

---

[7]The most explicit and thoroughgoing exponent of the view that social inquiry needs to be pursued as the science and technology of human manipulation is of course Skinner (1973). In a sense we all ought to be grateful to Skinner, for he simply makes sharply explicit assumptions that are implicit in a great deal of social science. Skinner, one might say, provides an unintended *reductio ad absurdum* of the whole programme of social inquiry pursued as social *science* and social *technology* on analogy with natural science and technology. It is most important in this connection to recognize that Skinner's general programme of developing a technology of human manipulation based on a predictive theoretical knowledge of human behaviour is quite distinct from the specific way in which Skinner proposes to implement this programme, namely in terms of behaviourism. The general

actions may be law governed). People are capable of discovering that they are being manipulated, and capable of disrupting manipulative predictions, whereas natural phenomena do not have such capacities. As a result, fortunately, manipulative social technology is not very effective, except in extreme circumstances, as in the case of torture or brain-washing.

Despite its inevitable ineffectiveness, the approach to social engineering that I have just described nevertheless has, I suggest, seriously damaging human repercussions. For in a thoroughly insidious fashion, it insinuates the idea that our fellow human beings are to be dealt with by means of manipulative techniques, this, furthermore, being the proper scientific, rational way to deal with other people. The idea that we might wish to join with our fellow human beings in worthwhile, valuable cooperative endeavours almost disappears.

There is a further consideration. Specifically psychological or social laws governing human, social action are only likely to be applicable as long as fixed, human, social aims are pursued in fixed, stereotyped fashions, by means of fixed methods. The moment active, human, social, intellectual inquiry exhibits a little more life, a little more innovation and creativity, pre-existing laws are likely to be disrupted. In the field of educational psychology, for example, laws governing learning and education are only likely to be successful as long as certain standard, fixed aims and methods continue to be put into practice. A more creative, aim-oriented rationalistic approach to learning and education would quickly render pre-existing 'knowledge' of educational psychology redundant. Likewise, a more aim-oriented, rationalistic approach to politics might quickly render pre-existing 'knowledge' of political science redundant. (For 'aim-oriented rationalism', see chapter 5.)

This reveals both the triviality, and the profoundly anti-humanitarian character, of the kind of theoretical knowledge

---

programme can survive intact even if behaviourism collapses. In opposing Skinner, it is the general programme that needs to be criticized, and replaced with something better; it does not suffice to refute behaviourism (even though, in my view, behaviourism deserves to be rejected as an absurdity, based as it is on an operationalist misunderstanding of the nature of physical science, which holds unobservable entities to be 'unscientific').

In exceptional cases, of course, manipulation of people may well be morally legitimate, as when the manipulated person cannot take full responsibility for himself due to being insane, mentally retarded, very young, or obsessional in some way, the person knowingly and voluntarily submitting to manipulative treatment, aversion therapy let us say, in order to gain greater self-control.

sought by social 'sciences' conceived of in accordance with the philosophy of knowledge. The social 'sciences' are only likely to meet with progress if there is no human progress. The more creative and innovative people are in their lives, the more rapidly will any 'laws' of the social 'sciences' become redundant. Far from helping to promote human progress and rational, humane, social inquiry, the social 'sciences', in order to meet with success, actually require people to be obediently incapable of innovative thought and action. We are being invited to conceive of ourselves as incapable of reason and creativity.

6 *There are undesirable consequences for all human endeavour due to the creation of general distrust of reason.* As a result of masquarading as rational inquiry when in fact, without this being realized, it exemplifies a profoundly damaging kind of irrationality, organized inquiry, pursued in accordance with the philosophy of knowledge, will tend to create entirely unwarranted general distrust of reason, thus causing further harm. There are two ways in which this will come about. First, general acceptance of the philosophy of knowledge as constituting 'rational thought' must tend to have the consequence that, when such inquiry leads to diverse undesirable consequences, of the kind discussed above, many people will conclude that reason itself is to blame, reason being somehow inherently defective, from a moral or human standpoint. Science and technology, making possible widespread suffering, injustice and death (see 2 above), even colluding with national and commercial actions that cause suffering, injustice and death (see 3 above) will be taken to demonstrate an inherent defect in the 'scientific' or rational approach to human problems. Second, as a result of reason becoming identified with the methods, the intellectual standards, of science as conceived of by the philosophy of knowledge, 'reason' becomes pecularly ill-equipped to help us tackle personal and social problems of living. For this irrational, philosophy-of-knowledge conception of reason requires that, in order to be rational, we must exclude all consideration of feelings, desires, aims, values, personal experi-ence and imagination, and attend only to impersonal, objective fact, evidence and logic. Even in the natural sciences such a conception of rationality is, as we have seen, irrational and damaging. But in most of the rest of life – and especially in connection with cooperative action and relationships of friendship and love – it is absolutely disastrous. The outcome of all this is that rationality comes to seem severely damaging if employed un-

reservedly in personal and social life, rationality apparently only being fruitful in limited domains such as natural science and mathematics. The programme of developing more rational ways of life, a more rational world, comes to be vehemently opposed in the interests of sanity, freedom, individuality, civilization. Precisely those who ought to be most concerned to help promote cooperative rationality in life – those who are most concerned to help create a more humane, a more just, a happier, more loving and wiser world – come to be the most vehemently opposed to it. Instead of rationality being understood to be essential for the flourishing of humanity, friendship, freedom, justice, love, civilization, it is seen to be, in many ways, the enemy of these things. (The mistake in all this, let me repeat, is to identify reason with the *irrationality* of the philosophy of knowledge. It is not reason itself that many self-confessed anti-rationalists oppose, but rather something that they have been fooled into taking to be reason, a characteristic kind of irrationality, long upheld by self-styled 'rationalists' to be reason itself.)

Opposition to science, reason, and the ideal of a rational world, based on the understandable misconceptions just indicated, is widespread and influential in the modern world. It is a major, enduring theme in literature, various expressions of which are to be found in the writings of Blake, Dostoevsky, Kierkegaard, Barzun, Ellul, Roszak, Zamyatin, D. H. Lawrence, Frisch, Laing, Cooper, Barratt, Feyerabend, and many others.

This concludes my survey of some of the ways in which socially influential inquiry, pursued in accordance with the philosophy of knowledge, must inevitably come to have damaging repercussions for all aspects of life as a result of the basic irrationality of the philosophy of knowledge.

CHAPTER FOUR

The Philosophy of Wisdom

The philosophy of wisdom is designed to overcome the fundamental and profoundly damaging defects of rationality inherent in
the philosophy of knowledge. It differs radically from the
philosophy of knowledge. All aspects of inquiry, all intellectual
disciplines and the way these are related to each other and to the
rest of society, are affected as we move from the philosphy of
knowledge to the philosophy of wisdom. There is, however,
nothing arbitrary about the basic principles of the philosophy of
wisdom. These principles, as set out below, are necessarily what
they are in order that the basic objective may be achieved: a kind
of inquiry that is devoted, in a genuinely rational way, to enabling
people to realize what is of most value to them in life.

Even though it has manifold repercussions, the basic idea of the
philosophy of knowledge is, as we have seen, extremely simple. It
is that inquiry can best help us realize what is of value in life by
devoting itself, in the first instance, to achieving the intellectual
aim of improving knowledge, in a way which is dissociated from
life and its problems, so that knowledge thus obtained may
subsequently be applied to helping us solve our problems of living.

It is just this basic, simple idea that the philosophy of wisdom
rejects as damagingly irrational. It holds instead that inquiry, in
order to be rational, in order to offer us rational help with realizing
what is of value, must give absolute intellectual priority to our life
and its problems, to the mystery of what is of value, actually and
potentially, in existence, and to the problems of how what is of
value is to be realized. Far from giving priority to problems of
knowledge, inquiry must, quite to the contrary, give absolute
priority to the intellectual tasks of articulating our problems of
living, proposing and criticizing possible solutions, possible and
actual human *actions*. The central and basic intellectual task of
rational inquiry, according to the philosophy of wisdom, is to help
us imbue our personal and social lives with vividly imagined and

criticized possible actions so that we may discover, and perform, where possible, those actions which enable us to realize[1] what is of value – happiness, health, sanity, beauty, friendship, love, freedom, justice, prosperity, joy, democracy, creative endeavour, cooperation and productive work – it being understood, of course, that knowledge and understanding can in themselves be of value in life, and are vital dimensions to almost all that is of value in life.

Far from serious, prestigious inquiry being primarily scientific or academic, it is according to the philosophy of wisdom, if anything, all the other way round: for each one of us the most important and fundamental inquiry is the thinking that we personally engage in (on our own or with others) in seeking to discover what is desirable in the circumstances of our life, and how it is to be realized. Institutionalized inquiry is simply a development of our personal and social thinking, having as its basic task to help us rationally develop our own personal and social thinking and problem-solving, so we may all the better realize what is of value to us in our personal and social lives. Whereas for the philosophy of knowledge the fundamental kind of rational learning is acquiring knowledge, for the philosophy of wisdom the fundamental kind of rational learning is learning how to live, learning how to see, to experience, to participate in and create what is of value in existence.

The central task of inquiry is to devote *reason* to the enhancement of *wisdom* – wisdom being understood here as the desire, the active endeavour, and the capacity to discover and achieve what is desirable and of value in life, both for oneself and for others. Wisdom includes knowledge and understanding but goes beyond them in also including: the desire and active striving for what is of value, the ability to see what is of value, actually and potentially, in the circumstances of life, the ability to experience value, the capacity to help realize what is of value for oneself and others, the capacity to help solve those problems of living that arise in connection with attempts to realize what is of value, the capacity to use and develop knowledge, technology and understanding as needed for the realization of value. Wisdom, like knowledge, can be conceived of, not only in personal terms, but also in institutional or social terms. We can thus interpret the philosophy of wisdom as asserting: the basic task of rational inquiry is to help us develop wiser ways of living, wiser institutions, customs and social relations, a wiser world.

---

[1]The phrase 'to realize what is of value' I use throughout to mean both 'to become aware of what is of value' and 'to make real or actual what is of value potentially'.

What ought we to mean by 'reason'? What is involved in tackling problems 'rationally'? Reason, according to the philosophy of wisdom, appeals to the idea that there are entirely general rules or methods of problem-solving which, when put into practice, other things being equal, give us our best chances of successfully solving our problems. It is essential to the conception of reason employed here that reason cannot, and is not intended to, dictate decisions to us. In acting and thinking in a wholly rational fashion we do not in any circumstances forego our freedom, or reduce freedom to the one free decision to act and think in accordance with the rules of reason: on the contrary, by acting and thinking in accordance with the rules of reason we maximize our freedom, our capacity to decide for ourselves *well*. (The basic task of reason is indeed to maximize freedom in the sense of freedom to achieve what is desirable and of value to us – all but identical to wisdom.) But there is of course nothing infallible about reason: however rationally we may act and think, we may still unnecessarily fail.

Two rules of rational problem-solving are absolutely basic: (1) articulate, and try to improve the articulation of, the problems to be solved; (2) imaginatively propose and critically assess possible solutions. In devoting reason to the enhancement of wisdom, academic inquiry gives absolute priority to these two rules of rational problem-solving.

Here we are, alive for a while, and then we die. How can we make something significant and of value our of our lives during the few decades that are, at most, available to us? How can we develop an ecologically sustainable world in which people do not die unnecessarily for lack of food, sanitation, medical care – a world in which there is a much more just distribution of land, resources, and wealth amongst people than at present? How can we put a stop to the nuclear arms race, the cold war, to the spread of armaments throughout the world, and to war both within and between nations? How can we get rid of dictatorships everywhere, and establish instead traditions of democracy and personal liberty? How can we develop more cooperative ways of working and acting together, so that ownership and responsibility are shared amongst adults, and so that our best, our noblest impulses may flourish?

These are the kind of problems, already referred to in chapters 1 and 3, that need to be put at the heart of the academic enterprise. These are the problems that academic inquiry should be centrally concerned to help us solve.

Not everyone, I imagine, will agree with my list of fundamental personal and social problems of living. For a number of reasons, what we take our problems to be is itself controversial and problematic.[2] Different people, different groups of people, encounter different problems. One person's solution may be another person's problem. Not all human failure and suffering constitute problems. We must recognize that some suffering is unavoidable, inherent to life. We all, at best, grow old and die. Problems arise when we suffer, when we fail to achieve what is desirable and of value, and our suffering, our failure could have been avoided. This makes the identification of our problems doubly problematic. In order to know what our problems are we need to know both what it is possible for us to do, and what it is genuinely desirable and of value for us to achieve.

It is just this inevitably *problematic* character of our problems which makes it essential for academic inquiry to devote considerable attention to the task of *improving the articulation* of our problems of living. The wide range of ways in which people, with different interests and beliefs, see their problems must be represented within academic inquiry, together with the cooperative endeavour to improve the formulation of these problems. It may seem that admitting such a plurality of interests and problems into the academic enterprise must inevitably destroy its coherence – so that it fragments into hostile, non-communicating factions. Such an outcome is possible: but it is not inevitable. At least there ought not to be any serious intellectual difficulty in establishing a common ground for the cooperative discussion of conflicting interests and problems within academic inquiry. Insofar as academic inquiry has, as its basic task, to devote reason to helping humanity achieve that which is of value in life, we can agree that inquiry must be committed to helping people resolve their problems in a cooperative and just fashion – to the extent that this is possible.

Solutions to personal, social problems of living are essentially personal, social *actions*. Thus, according to the philosophy of wisdom, academic inquiry is centrally and fundamentally con-

---

[2]There is, for example, a considerable difference in the way the Brandt Commission saw global problems of economic development, and the way some of its critics understand such problems. See Brandt *et al.* (1980) and *Encounter* (1980). Richard Barnet (1972) has analysed brilliantly the different ways in which United States administrations and third-world revolutionary movements perceive and understand problems of the third world. The inevitably problematic character of problems is perhaps the main consideration which leads me to develop aim-oriented rationality in chapter 5.

cerned to propose and assess critically possible and actual personal, social actions, from the standpoint of their capacity to help us achieve what is of value in life. The task of proposing and criticizing possible actions is actually and intellectually more fundamental than the task of proposing and criticizing claims to knowledge.

There are, of course, many rules of rational problem-solving in addition to the two basic rules already mentioned.[3] In tackling a complex problem it is often helpful to break the given problem up into a number of subordinate, specialized problems, which we tackle one by one, the solutions then being put together to solve our original, overall problem. It may be helpful to begin by tackling easier, analogous problems in an attempt to develop helpful methods of attack. In order to develop good ideas for a solution to our given problem it is often helpful to look at solutions to analogous, already solved problems.[4] Quite generally, in fact, solving a new problem involves discovering how to relate the new problem to analogous, already solved problems.[5] As a result of

[3]The best book, to my knowledge, on rational problem-solving is Polya (1957); it is also one of the simplest and most delightful. Polya is concerned with how to go about solving elementary mathematical problems: he makes it clear however that strategies that arise in connection with solving mathematical problems are relevant to discovery and problem-solving in general. Also of interest in this context are Hadamard (1954) and Lakatos (1976). De Bono's tireless efforts to promote practice in, and a sense of the importance of, problem-solving also deserve to be mentioned. See, for example, de Bono (1972, 1974).

[4]Given that we seek to solve a problem P, and that a different but vaguely analogous problem P_1 has a known solution S_1, we may seek gradually to modify P_1 in the direction of P, at the same time appropriately modifying S_1, so as to be a solution to the modified problem until eventually it becomes a solution to P itself.

[5]An important additional rule is: try reformulating the problem P to be solved, and try reformulating the reformulation, and so on, in this way building up a network of reformulated versions of P, any one of which, if solved, leads to a solution to P (immediately, or without too much difficulty) in this way endeavouring to arrive at a solvable distant cousin of P. This rule helps to explain why problem-solving may actually be a more methodical, less irrational process than it is often thought to be. In support of problem-solving being irrational it is sometimes argued that solutions often come in a flash, in a moment of inspiration, almost unsought, often when methodical searches for a solution have persistently failed. What the above rule suggests is that this common phenomenon may well be deceptive. The result of applying this rule methodically and laboriously to some problem P may be the discovery that if P_N can be solved, so can P. In a flash it may be recognized that P_N is easy to solve. The high excitement of at last discovering how to solve P may fool one into supposing that the discovery that P_N can be solved is a moment of inspiration, of high intellectual achievement. Actually it may be nothing of the kind. P_N may be genuinely very easy to solve. The achievement lies in the laborious methodical discovery that a solution to P_N enables one to solve P.

putting into practice these kinds of additional rules of rational problem-solving, we develop a *tradition* of problem-solving, which enables us progressively to build up, to enhance our problem-solving power. Rational problem solving involves quite essentially the progressive development of problem-solving power in this way.[6]

All this, according to the philosophy of wisdom, is exploited by academic inquiry. The basic task of academic inquiry is to help us build rules such as these into our habits of thought, feeling and action, into our personal and social life, and into our institutions, so that we may tackle our problems of living in such a way as to give ourselves the best chance of realizing what is of value to us, thus progressively enhancing our powers to realize value in life – progressively enhancing our freedom, our creativity, our capacity to love, our wisdom. In particular of course, academic inquiry itself puts these rules into practice, in rationally searching for solutions to problems of living. This does not just involve individual scientists and scholars putting these rules into practice

---

Thus appearances to the contrary, all the really difficult and substantial work involved in discovering how to solve P was actually performed in a slow, progressive highly methodical way. (This point arises in connection with other rules as well.)

Two further rules of rational problem-solving ought perhaps to be mentioned. First, in attempting to solve any given problem, always be ready to *change the problem*. The given problem P may be unsolvable, and may need to be changed to P_1, the most desirable solvable problem close to P. Alternatively it may be undesirable to solve the given problem P despite first appearances to the contrary: it may be desirable to change P to P_1, a desirable, solvable problem close to P. Second, in attempting to solve P, always be ready to consider, and reconsider, P as subordinate to some larger, more general, or more fundamental problem P_1, the solution to P being sought in order to help solve P_1. New approaches to solving P_1 may require P to be modified; or may render it unnecessary, or even undesirable, to solve P (in which case persisting in the attempt to solve P becomes irrational). These two rules are basic to aim-oriented rationality, to be expounded in the next chapter. It may be noted that, from the standpoint of the philosophy of wisdom, the philosophy of knowledge is irrational because it prohibits inquiry from putting these two rules into practice. We seek to solve problems of knowledge because, more fundamentally, we seek to solve problems of living. The two rules just indicated require that attempts to solve problems of knowledge be rationally responsive to attempts to solve more fundamental problems of living. This, demanded by the philosophy of wisdom, is prohibited by the philosophy of knowledge, in a misguided attempt to preserve the 'objectivity', the 'rationality' of science.

[6]These rules of reason presuppose that we can already successfully solve problems in the world; they are designed merely to help us marshal our already existing problem-solving power in order to solve new problems. All that reason can accomplish is to help us to reorganize what we can already do – solutions to

in their own individual research work; in addition it involves these rules being built into the whole intellectual/institutional structure of the academic enterprise – thus influencing such things as the way that disciplines are related to each other and to the human world beyond; decisions concerning what is to be published; decisions concerning what research is to receive financial support; academic appointments; the content and style of education – of seminars and lectures, degree courses, examinations. Emerging out of, and feeding into, the central concern with our personal and social problems of living, academic inquiry quite properly creates and explores a wide range of subordinate, specialized intellectual problems, academic work on these subordinate problems all being designed, in one way or another, to help us achieve what is of value in life. Thus the *technological sciences – engineering, medicine, artificial intelligence* – seek to solve, and to develop techniques for solving, those technical problems that need to be solved if we are to realize desirable life-aims such as prosperity, health, release from repetitive, soul-destroying work. *Mathematics* seeks to develop, systematize and unify abstract problem-solving methods, applicable to as wide a range of circumstances as possible. Pure mathematics is concerned with significant, problematic possibilities, and not with anything actual at all. The *physical* and *biological* sciences seek to solve subordinate problems of knowledge and understanding concerning diverse aspects of the natural world. The *humanities* and the diverse branches of *social*

---

problems that we can already solve – so that they become a solution to the new problem P that we initially do not know how to solve. This point has an important bearing on a basic tenet of the philosophy of wisdom (to be discussed below) that successful action in the world comes before, and is presupposed by, thought, reason and knowledge. The point is also important in connection with the Humean problem of rational action – the problem of how there can be any such thing as rational action in the world. And finally the point explains why it is important, of such value, to tackle our problems rationally: in doing so, we give ourselves, other things being equal, the best chances of progressively enhancing our problem-solving powers. According to this view, in acting rationally we act in such a way as to give ourselves the best chances of successfully developing and extending what we can already do. The basic task of reason is to help us to establish traditions of learning, of making progress.

This immensely important point – that in a sense we only ever improve on what we can already do – might even be enshrined in another rule: in learning how to do something entirely new, which you cannot at present do at all, begin by *doing it* (in some non-destructive way) and then set about progressively improving your performance. It is this rule that young children put into practice so successfully in learning to speak: first they babble, then gradually transform this babbling into speech.

inquiry have the fundamental intellectual task of articulating our problems of living, proposing and critically assessing possible solutions. In sharp contrast to the philosophy of knowledge, the philosophy of wisdom holds social inquiry to be intellectually more fundamental than the natural sciences, just because social inquiry is concerned with primary problems of living whereas natural science is concerned with subordinate and secondary problems of knowledge. (From the standpoint of the philosophy of wisdom, the proper term is social *inquiry* rather than social *science* – the latter being a typical philosophy-of-knowledge misnomer and misconception.) Insofar as the humanities and the diverse branches of social inquiry seek to improve our knowledge and understanding of people and societies, this is undertaken as a subordinate intellectual enterprise in order to aid the fundamental task of helping us to realize what is of value in life. Thus *economics* has as its primary task to propose and criticize possible solutions to economic problems – the economic aspects of our problems of living – contributions to economic theory and knowledge being intellectually subordinate and secondary. *History* and *anthropology* have, as their basic tasks, to acquaint us with the successes and failures that people have encountered in seeking what is of value in life in the past and in other places, in other cultures, so that we may learn from their example. Keeping a record of our past problem-solving efforts is essential to reason. History and anthropology thus make an essential contribution to reason. *Psychology* has, as its primary task, to help us to articulate our personal and interpersonal problems of living, and to propose and criticize possible solutions, as we live, thus helping us to resolve rationally our most personal, emotional, intimate problems. *Sociology* has, as its basic task, to help us to propose and criticize possible solutions to social and institutional problems, in relevant social and institutional contexts, so that we may gradually improve our capacity to resolve such problems in a cooperatively rational way. *Political inquiry* has, as its basic task, to help us to articulate our diverse political problems, and problems associated with government, and to propose and criticize possible solutions, thus helping us gradually to discover how we may tackle these problems in a more cooperatively rational way than we do at present. The *study of international relations* has, as its basic task, to help humanity to articulate its international, global problems, and to propose and criticize possible solutions, so that gradually we may discover how to resolve these problems in a more humane, just, cooperatively rational way than we do at present. Yet again, *philosophy* seeks to articulate and assess critically views as to what ought to be the aims and methods of our diverse pursuits – art, literature, politics, theatre, education, industry, commerce, law, science. Philosophy is thus a severely practical endeavour: its task is to help us improve the aims and methods of our various pursuits, as we act. It is 'philosophy' in just this sense that I am engaging in here, in expounding and critically assessing two rival views about what ought to be the basic aims and methods of the academic enterprise, the 'philosophies' of knowledge and wisdom.

The primary intellectual aim of the humanities and social inquiry, quite generally, is to help us to realize what is of value to us in our personal and social lives. What ultimately matters is personal and social progress towards enlightenment and wisdom: all academic progress is but a means to this end.

Academic inquiry may develop an intricate maze of subordinate, specialized academic disciplines and problems: it is however vital – according to the philosophy of wisdom – that work within these specialized disciplines on these specialized problems be undertaken in such a way that the overall outcome of, and reason for, this work is our enhanced capacity to solve our fundamental problems of living. In other words, specialized problems must be tackled as rationally subordinate to our fundamental problems of living.[7] Only in this case can academic inquiry hope to be rational, intellectually rigorous and of maximum human value. The intellectual progress and success of academic inquiry is to be judged in terms of the extent to which academic work produces and makes available ideas, proposals, arguments, discoveries, techniques that help people achieve what is of value in life in a cooperative and just way.

The fact that scientific problems are tackled as aspects of, and in a rationally subordinate way to, intellectually more basic personal and social problems of living does not mean that only those scientific problems ought to be tackled whose solutions have immediate and obvious technological applications. There are two reasons for this. First, science can contribute to the quality of life, to the enhancement of wisdom, directly by enhancing our knowledge and understanding of significant aspects of the world around us (such knowledge, perhaps, having no technological applications). Second, problems are often solved in an unexpected, indirect way, as an unforeseen consequence of a solution to an apparently unrelated problem – a familiar phenomenon in both

---

[7]This point is developed further in Maxwell (1980).

science and mathematics. If it was always obvious what scientific research programmes need to be pursued in order best to help us solve social problems of living, there would be little point in stressing the need to interrelate imaginative and critical discussion of social and scientific problems. It is precisely because this point is not obvious that the interrelation between social and scientific problems needs sustained, explicit, critical attention.

According to the philosophy of wisdom, all the intellectual problems and aims of all science and scholarship are fundamentally personal and social in character. This does not mean, however, that the only kind of value that inquiry is recognized to have is a practical value. Quite to the contrary, the philosophy of wisdom seeks to emphasize the profound value that inquiry can have when pursued for its own sake, and not only as a means to some other end. Realization of value (the aim of all inquiry) includes the seeing, appreciation and understanding of what is of value, in people, in art, in the world, as well as the active endeavour to cherish and help grow what is of value, potentially and actually, in existence. The philosophy of wisdom insists, however, on the profoundly *personal* and *inter-personal* (or social) character of inquiry pursued for its own sake. Our own personal endeavour to see, to understand, what is of value in existence as we live is, for each one of us, pure inquiry, inquiry pursued for its own sake, at its most fundamental and important. In order to appreciate just how precious we hold such personal inquiry to be, consider why we would be so distressed to discover we are about to become blind. In part this distress would be due to the prospect of being deprived of the practical value of seeing; but far outweighing this, surely, would be the distress that we would feel at the prospect of being deprived of sight for its own sake. A whole precious dimension of personal inquiry – discovering and experiencing the visual aspect of things – would be cancelled. The extent of the distress we would experience at the prospect of being deprived of sight for its own sake gives us an indication of how highly we value our own personal inquiry pursued for its own sake. (We are all perhaps inclined to devalue, even to ignore, our own personal participation in inquiry pursued for its own sake because we tend to identify such inquiry with expert academic inquiry, our own thinking being depreciated as a result for failing to comply with the intellectual standards of the philosophy of knowledge. This is one way in which the philosophy of knowledge harms inquiry pursued for its own sake.)

According to the philosophy of wisdom, the whole *raison d'être* of academic inquiry, from a purely intellectual standpoint, is to promote and aid personal inquiry, pursued for its own sake, as an integral part of life. According to this view, even an academic discipline as apparently remote from human concerns as *cosmology*, has a profoundly personal, social and creative aim: to enable people to improve their own personal knowledge and understanding of this cosmos in which we live.

As the argument of this book unfolds, I shall be concerned to stress that the philosophy of wisdom does better justice to *both* the practical *and* the intellectual aspects of inquiry than does the philosophy of knowledge. Thus, the philosophy of wisdom stresses the intellectually fundamental character of articulating problems of living, proposing and criticizing possible solutions. At first sight this has only a practical value. I shall argue, however, that such imaginative exploration of people's problems is precisely what we need to do in order to acquire what may be called person-to-person understanding of other people – a kind of understanding (promoted by great literature) that it is essential to acquire if we are to appreciate what is of value in other people's lives, the value-discoveries of others enriching our own. This kind of person-to-person understanding is, I shall argue, fundamental to our humanity, essential for reason, and even for science. And yet the philosophy of knowledge debars it from rational inquiry for failing to satisfy its (misconceived) intellectual standards of 'objectivity' and impersonality. Yet again, much actual academic inquiry ostensibly pursued only for its own sake, may actually be pursued for quite different, all-too-human reasons: to further academic careers or win fame or status. As we shall see, the intellectual standards of the philosophy of wisdom can help us put right such perversions of science and scholarship – whereas those of the philosophy of knowledge cannot help.

The transition from the philosophy of knowledge to wisdom changes dramatically the whole way in which the two aspects of inquiry – 'pure' and 'applied' – are conceived. From the standpoint of the philosophy of knowledge, the two aims of inquiry (knowledge for its own sake, for the sake of its technological applications) seem to be quite distinct, even if a contribution to knowledge may be of value in both ways. From the standpoint of the philosophy of wisdom, the two aims ought to be intimately interrelated. In pursuing inquiry for its own sake we seek to discover what is of most value in existence. Even a meagre appreciation of what is of most value in existence can scarcely be had without some awareness of just how terrible it is that people –

millions of people – should be needlessly deprived of their one opportunity to experience and participate in what is of value. Thus there develops the active concern to help people resolve their practical problems of living. The motive for this concern, however, ought to be to make it possible for more people to enjoy what is of value in life for its own sake. The two aims of inquiry are united in love.[8]

The transition from the philosophy of knowledge to the philosophy of wisdom changes dramatically the relationship between academic inquiry and politics. According to the philosophy of knowledge, political programmes and problems ought to have no place in the intellectual domain of inquiry, which is concerned only with questions of fact and knowledge. According to the philosophy of wisdom, the intellectual domain of inquiry is concerned fundamentally with political programmes and problems. The distinction between the two spheres is based, not on subject matter, but on aims and methods. Academic inquiry is concerned to promote imaginative, critical thought, rational, cooperative, political action: it is not concerned to wield power, to legislate, or to persuade and manipulate, as are many of those who engage in political activity.

Again, the transition from the philosophy of knowledge to the philosophy of wisdom changes dramatically the relationship between science (or academic inquiry) and religion. According to the philosophy of knowledge, religious ideas and problems have no place within the intellectual domain of inquiry. According to the philosophy of wisdom, academic inquiry is concerned fundamentally with religious ideas and problems. If 'religion' is characterised in a broad way as 'concern for what is of most value in existence' then academic inquiry, as construed by the philosophy of wisdom, is essentially a religious enterprise. If 'God' is characterized in a sufficiently open, unrestricted way – as it ought to be according to many religious traditions – as that unknown something that is of supreme value in existence, then inquiry, as conceived of by the philosophy of wisdom, has as its overall goal to help us to realize 'God'. Inquiry as conceived of by the philosophy of wisdom is, however, opposed to the authoritarian and anti-rationalist elements present in most world religions. In particular, the idea that 'God' can be a supreme person, all powerful, all knowing and all loving, is rejected as a logical, moral and religious obscenity. Such a God would be knowingly responsible for all

[8]This point is developed further in Maxwell (1976b).

human suffering and death engendered by natural causes, and a participant in all suffering and death caused by people (since this invariably requires collaboration from Nature). Such a God would be a torturer and murderer of all mankind – infinitely more criminal than a mere Hitler or Stalin. All traditional attempts to excuse God's torturing and killing of people are similar to, and are on the same intellectual and moral level as, attempts to excuse the torturing and killing perpetrated by a Hitler or Stalin. To call such a cosmic tyrant a being of love is the most blatant inconsistency imaginable (unless one has monstrously perverted ideas about love). To advocate publicly that an all-powerful, knowing and loving God exists, as if this is a consistent possibility (let alone a known certainty) is, from the standpoint of the philosophy of wisdom, profoundly damaging in that it strengthens the impression that reason does not apply where it most needs to be applied: to the problem of what is of supreme value to us in existence. Where it is most important for us to be rational we become carelessly and destructively irrational.

The mistake is to identify *power* and *love*. 'God' in the sense of cosmic *power* is the unified pattern of physical law that runs through all phenomena, to be discussed in chapter 9: it knows nothing of human suffering and cannot love. 'God' in the sense of cosmic *love* is that which is best, most loving, potentially, in human life, to be discussed in chapter 10: though potentially profoundly loving, it is at present often both ignorant and powerless – and hence the need for inquiry pursued in accordance with the philosophy of wisdom.

It is not hard to understand why there should be such a great temptation to believe in the blatant inconsistency, a God of love who tortures and murders. We want our all loving God to be all powerful and knowing because we want to be assured that God can care for us, resurrect us after death, put to right all wrong. But in order to be comforted in this way we must abandon reason. Authentic religion seeks to help us confront realities however disturbing: it does not seek to console us with comforting illusion.

The transition from the philosophy of knowledge to wisdom also changes dramatically the relationship between academic inquiry and art. According to the philosophy of knowledge, art itself, like politics and religion, has no rational place within academic inquiry – even though of course factual knowledge about art, politics and religion does have such a place. According to the philosophy of wisdom, literature, drama, music, dance, painting, sculpture and other forms of art can make major and fundamental *rational* contributions to inquiry – as revelations of value in the world, and as imaginative explorations of life-problems and their possible resolution. Ancient Greece and the Italian Rennaissance provide striking illustrations of how fundamental the contribution of art can be to inquiry if the latter is not hermetically sealed off from such learning as a result of observing the edicts of the philosophy of knowledge.

The two philosophies have radically different implications for education. Academic inquiry shaped by the philosophy of knowledge inevitably leads to education being of two kinds, often at odds with each other. On the one hand there is academic learning; on the other hand there is learning about how to live. If academic inquiry is shaped by the philosophy of wisdom, this dichotomy disappears. Academic learning *is* then learning about how to live. The philosophy of wisdom intelligently put into practice in schools and universities would change education beyond all recognition. Many current conflicts, difficulties, failings, would disappear. All education would be what children instinctively want it to be: learning about how to live, learning about how to realize what is of most value to us in the circumstances of our lives.

The two philosophies uphold requirements for rationality, for intellectual rigour, that are in important respects diametrically opposed. Thus, far from it being necessary for inquiry to be dissociated from life and its problems in order to be rational, it is, according to the philosophy of wisdom, all the other way round: inquiry can only rationally and effectively perform its basic task of helping us realize what is of value insofar as it is an integral part of our lives – even academic and scientific inquiry needing to be in close contact and communication with persons and institutions in the non-academic, non-scientific world in order to be able rationally to aid the realization of value in life. Again, far from it being necessary to banish desires and feelings from the intellectual domain of inquiry in order to preserve its rationality, it is all the other way round: desires and feelings must form an integral part of the intellectual domain of inquiry, at the most fundamental level (our own personal thinking) if inquiry is to be rational – capable, that is, of achieving rationally or effectively its basic task. Not everything that feels good is good, and not everything that we desire is desirable: but devoid of our feelings and desires we can make no value discoveries of our own: we can but echo or mimic the value discoveries and achievements of others. Thus, if inquiry is to help us realize what is of value, it must attend to our feelings and desires: the very articulation of our problems of living requires the expression of feelings and desires. According to the philosophy of wisdom, in fact, reason – rational action – is essentially so interrelating action, experience, feeling, desire, aim, imagination and doubt that we give ourselves the best chances, other things being equal, of realizing what is of value. Only by bringing together desires, aims, feelings, deeds and objective facts imaginatively and critically can we hope to be rational, and come to appreciate something of the value of what there is in the world. Whereas the philosophy of knowledge seeks to shield inquiry from an irrational world in order to preserve intact its rationality, the philosophy of wisdom, by contrast, gives to inquiry the basic task of helping us gradually develop more rational lives, a more cooperatively rational human world.

One assumption that tends to lie behind the philosophy of knowledge is that rational action only becomes possible once relevant knowledge has been obtained. This assumption is rejected absolutely by the philosophy of wisdom. What is absolutely fundamental is life itself, our doing things more or less successfully in the world, and our capacity so to do things. Our lives, our actions, are rational to the extent that we are able to exploit to our best advantage what we can already do in order to do new things so as to solve new problems. Being able to imagine possible actions can enormously increase our rational problem-solving power – if only because of the advantages to be accrued from trying out diverse actions in our imagination only, and not in the real world. Propositional knowledge and science are but developments of these more fundamental capacities, explorations in effect of what we can and cannot do, actually or in principle.

According to the philosophy of wisdom, the physical and biological sciences have an enormously important role to play within academic inquiry as a whole. Knowledge and understanding of the natural world are vital dimensions of wisdom. The crucial point, however, is that if the scientific search for knowledge and understanding is to be undertaken rationally, within inquiry as a whole, then it is essential that it be rationally subordinated to the intellectually more fundamental search for value in life (problems of knowledge being rationally subordinated to problems of living, natural science being rationally subordinated to social inquiry).

None of this means, let it be noted, that accepting the philosophy of wisdom rather than the philosophy of knowledge leads to a greater tendency to accept as true that which is highly desirable if true, or to reject as false that which is highly undesirable if true. Quite the contrary, as a result of taking human desires and aims into account in assessing contributions to knowledge, as the philosophy of wisdom requires, we put ourselves in a better position to correct any tendency to suppose that desirability implies truth, undesirability falsehood. The philosophy of wisdom provides us with a more intellectually rigorous conception of science, and of inquiry, than does the philosophy of knowledge, and upholds a more exacting, and more widely applicable, conception of reason. Stalin's imposition of Lamarkism on Soviet biology is no more in accordance with the philosophy of wisdom than it is in accordance with the philosophy of knowledge. Indeed, this argument for the philosophy of knowledge badly backfires: whereas philosophy-of-knowledge science now flourishes in the Soviet Union, despite lack of free speech, philosophy-of-wisdom science would not be tolerated. This indicates how restricted, how tame, philosophy-of-knowledge intellectual standards really are. (It should be noted, in addition, that the philosophy of wisdom fully recognizes the elementary point that valuable contributions to inquiry can be made by those who pursue bad aims, and that trivial or harmful contributions can be made by those who pursue good aims.)

We live, it seems, in an impersonal universe. Insofar as there is anything of value in the universe it has to do with life and especially, for us, with our own lives here on earth, the way in which we are in this impersonal cosmos. It is this holy mystery, this miracle of our existence with all its potentialities, embedded in the cosmos, that rational inquiry has as its task, its charge, to help us to cherish and grow, in an adult and responsible fashion, so that what is of value in our lives flourishes. Awareness of our surroundings and of ourselves is certainly a part of what is of value: but in order to be fully of value this awareness needs to inform our lives, our deeds, and not be cut off from life as impersonal scientific knowledge. The supreme thing, perhaps, is to live life lovingly, insofar as we can, lovingness certainly including every attention to the reality of what is objectively of value in the world, in others, in oneself. It is just such an objectively loving way of life that the intellectual standards of the philosophy of wisdom can promote, and that the intellectual standards of the philosophy of knowledge must sabotage.

Inquiry as conceived of by the philosophy of wisdom is perhaps best understood as being similar in character to, even though a rational development of, animal inquiry, animal learning. Animal learning is learning how to act, how to do, how to live. It is precisely in this way that we need to see our finest ideal of rational human learning, of rational human inquiry. A major difference is that whereas animals learn how to act to discover how to survive and reproduce, we may demand that we learn how to act in order to discover how to realize additional goals of value to us, such as justice, democracy, understanding, friendship and love. There is for us the possibility that we can exercise some influence over the quality of what survives and is reproduced – over what we can become and be and what our children can become and be.

It should be emphasized that, from the perspective of the philosophy of wisdom, all the diverse defects of the philosophy of knowledge follow from one single but fundamental error: a profound and disastrous misrepresentation of the basic intellectual aim of inquiry. All the other defects of the philosophy of knowledge – its failure to characterize adequately such things as the proper relationship between inquiry and life, the aims and methods of social inquiry, the aims and methods of the natural sciences, the proper relationship between social inquiry and natural science, the nature of intellectual progress, the place of human desires and feelings in inquiry and their relevance for rationality, and so on – all these diverse failings stem from the simple, basic failure to specify the proper overall intellectual aim of inquiry.

A major task of this book is indeed to get across *both* (1) a sense of just how simple, how elementary, the basic proposal is – to change the overall aims and methods of inquiry, from knowledge to wisdom, in order to enhance simultaneously the rationality and the potential human value of inquiry; *and* (2) a sense of just how diverse and wide-ranging the repercussions of this simple proposal are – to such an extent that were we to take the proposal seriously, no aspect of our personal, social, political, intellectual or cultural life would remain unaffected.

The simple, elementary character of the basic proposal is exhibited in figure 3.

It is of course not the job of scientists and scholars actually to decide for the rest of us what our problems are, how they should be solved, what we should do with our lives, and what is of value. On the contrary, the proper job of scientists and scholars is to help the rest of us to reach our own decisions – decisions that we do really want to make. In other words, the proper task of academic inquiry is not to deprive us of our power to choose and decide but to enable us to enhance our power to choose and decide well. The most important and fundamental kind of thought that there is in the world exists as an integral part of our personal and social life.

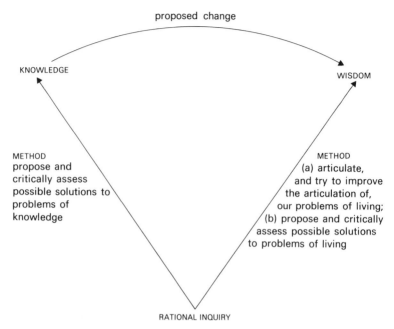

Figure 3 Proposed change in aims and methods of inquiry

Subordinate, specialized aspects of thought are *delegated* to academics, our servants not our masters. At the most fundamental level of all, there is life, actions, the ability to do things. Out of this there emerges conscious thought, imaginative problem-solving, as an integral part of personal and social life. This in turn gives rise to institutionalized, professional thought – science and scholarship – devoted in turn to helping to promote rational, enlightened thinking in life.

The way in which life, and personal and social thought, ought to be related to academic thought – according to the philosophy of wisdom – can perhaps be illuminated by considering an analogy drawn from music. Associated with music there are all sorts of highly specialized skills and fields of expertise, to do with such things as playing and making musical instruments, musical composition, musicology, teaching: in the end, however, the whole rationale for engaging in all such specialized, technical pursuits is to further the creation, enjoyment and appreciation of performed music. In an analogous way, and more generally, we may argue, the whole rationale for engaging in the diverse specialized, technical pursuits of academic inquiry is to further the creation, enjoyment and appreciation of value in life.

From the standpoint of the philosophy of wisdom, public organized inquiry is perhaps best understood as arising primarily in response to – and to help us to solve – the problems of acting cooperatively in a vast, complex, diverse, interconnected human world of the kind we live in today. When humanity lived in small hunting and gathering tribes, this problem did not exist. It is at

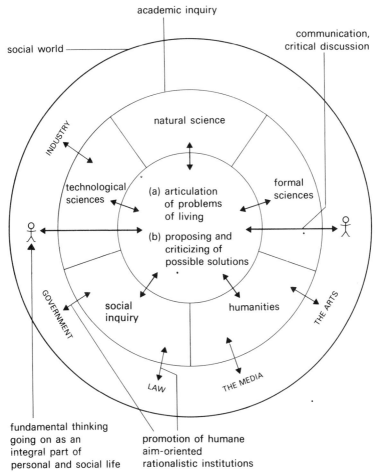

Figure 4 Relationship between the intellectual domain of inquiry and the social world according to the philosophy of wisdom

least possible for a tribe of some fifty people, who all speak the same language and share the same culture, skills and values, to solve problems cooperatively, without any elaborate institutional organization. Informal tribal meetings can be held to decide matters of concern to all, with everyone being able to have their say without major logistic problems being encountered. But then the hunting and gathering way of life gives way to more settled agricultural ways of life; tribes grow in size and begin to trade with neighbouring tribes; cities are built, tribes coalesce; divisions of class, work, skills, culture and values grow within societies of increasing size; modern science, technology and industry develop, and with them the further development of diverse, vital esoteric skills and expertise. Modern methods of communication, travel and trade have the effect of interconnecting most people, at least to some extent, to form a vast, complex interdependent and interacting global society. A tribal meeting of humanity has become a logistic impossibility – thus making cooperative action extraordinarily difficult to achieve. (Cooperative action is here to be understood to imply action engaged in by a number of people who share responsibility for what is done and for deciding what shall be done, to their general benefit, there being no permanent leadership, or delegation of responsibility.)[9] Cooperative action – whether performed by groups of friends or colleagues, local communities or at the national or international level – is only possible if there exists the means for those acting to discuss problems, objectives and diverse possible actions. The most basic and urgent task of public, organized inquiry – academic inquiry – is, according to the philosophy of wisdom, to provide the means for such discussion ultimately at the global level – thus providing a sort of intellectual/institutional substitute for a tribal meeting of humanity. Intellectual standards governing the articulation and discussion of problems of living need to serve this end; standards of clarity, simplicity, truthfulness, justice, and cooperation. Contributions to academic inquiry – articulations of problems, proposals for action intended to solve human problems – must be assessed on their merit and not merely in terms of the expertise, authority or power that a person or group of people making the contribution may possess, or may be held to possess. The ability of people with power, wealth or talent to influence the way problems

---

[9]Cooperativeness may include competitiveness, if generally deemed to be desirable: it is not automatically the opposite of cooperation. Force, threat of force, manipulation: these are what negate cooperation.

of living are discussed in public, in the media and elsewhere, must be counteracted by academic inquiry so that the interests and problems of the powerless, the poor and the inarticulate receive their due attention and representation, and are not neglected. On this view, academe is a sort of people's civil service, doing openly for people what the civil service does for government.

In order to perform this task properly it may be necessary for academic inquiry to develop esoteric discussion and vocabulary – in natural science and mathematics – impenetrable to most people; it is also absolutely essential, however, that the most important problems and discoveries be formulated clearly, simply and non-technically, so that even twelve-year-old children can understand them.

In recent years Popper has placed great emphasis on the importance of recognizing the existence of a relatively auto-nomous 'World 3' of ideas, propositions, theories, problems and arguments, which interacts with the material 'World 1' via the psychological 'World 2' (Popper, 1972; Popper and Eccles, 1977). From the standpoint of the philosophy of wisdom, this Popperian view is a philosophy-of-knowledge mystification of the intellectual domain of inquiry. In reality there are, in this one world, the deeds, imaginings, proposals, intentions, arguments, suggestions and convictions of *people* – all aspects of personal and social life. It is, however, of vital importance that we develop in our human world a tradition of treating possible actions, proposals for action, problems, arguments, philosophies, theories, *as if* they are entities existing more or less independently of who expresses them, for what motive, in what language, when and where. We need to do this so that we may make possible something like a rational, cooperative tribal discussion of humanity, which cuts across barriers of language, time and place, and gives due emphasis to what is best, irrespective of the dictates of mere power or wealth.

If this intellectual domain of discussion is to perform its proper function, however, it is vital that we see it as a human fiction, created for a vital human purpose, an immensely valuable part of the social fabric designed to make possible just, cooperative action: it is vital that we do not become so dazed and inflated by this social creation of ours that we begin to imagine, with Popper, that it amounts to an autonomous realm of being.

A basic task of inquiry, according to the philosophy of wisdom, is to help us live more rationally in all that we do, and to help us develop a more rational social world. I have argued that many who apparently oppose reason, actually oppose its opposite – a characteristic kind of extremely damaging irrationality, associated with the philosophy of knowledge, which falsely masquarades as reason. Many I am sure will not be entirely convinced by this argument: they will continue to find the very idea of living a wholly rational life in a wholly rational world thoroughly objectionable.[10] I therefore now add a few remarks intended to make clear that it is always desirable to be rational in all that we do.

The central point is simply this. Acting and thinking in a wholly rational way cannot possibly in general go against our best interests, whatever we are doing, just because to act and think in this way is simply to do so in accordance with certain general methods or strategies which, other things being equal, give us our best chance of solving our problems, realizing what is of value to us. If reason leads us systematically astray then, by definition, it cannot be reason.

Consider the way in which attitudes towards reason have come down to us from the past. If we take what reason meant to Descartes, Leibniz, Hobbes or Spinoza as our starting point, and consider the subsequent development of the concept, then two kinds of decline in the concept become apparent. There is first of all a decline in what is taken to be the power, the efficacy of reason. For seventeenth-century Rationalists, reason had almost unlimited power in that, if properly used, it could decide issues with absolute certainty, beyond all doubt. Gradually since then this extreme confidence in the power of reason has been eroded. Sceptical arguments such as those of Hume led Rationalists to conclude reluctantly that reason may only be able to function negatively, in establishing the falsity of some claims to knowledge, finally even this negative power being called into doubt as the view develops that experimental refutations of theories in science are rarely decisive. Even the power of reason to establish results beyond all doubt in mathematics is called into doubt, as the paradoxes of set theory are discovered, introducing uncertainty even into the foundations of mathematics, the notion of rigorous

[10]This is a major theme of romantic literature, art, thought, politics, psychiatry and education, to be found in the writings of Blake, Rousseau, Dostoevsky, Nietzsche, Kierkegaard, D. H. Lawrence. (Russell gives a hilarious account of Lawrence's attempt to convince him to throw away reason and the intellect; see Russell, 1956, pp. 106–7.) For a delightful fictional portrayal of the horrors of the perfectly 'rational' society see Zamyatin (1972). For a novel portraying the breakdown of a 'rational' individual, see Frisch (1974).

proof becoming uncertain. The development of intuitionist logic, which rejects the law of excluded middle 'p or not p' indicates that uncertainty has even reached the inner sanctum of reason, logic.

The second kind of decline that reason has suffered since Hobbes and Spinoza is a decline in what is taken to be the proper domain of application, the scope, of reason. For Hobbes and Spinoza, reason seems to have unlimited scope in that it is applicable not just to questions of fact and knowledge, but also to moral issues, to problems of politics and religion. Subsequently, in part as a result of the development of the Romantic movement, this has seemed to many to be far too wide an application of reason, impossible to carry out, or undesirable to carry out even if possible. Thus religious faith is placed beyond the reach of reason, morality is held to be an affair of the heart and not of the mind, not something to which reason can be applied. Art, love, friendship, enjoyment, happiness may all be held to be beyond the domain of reason. The perfectly rational life or perfectly rational society become not ideals to be striven for, but nightmares, ultimate horrors, to be adamantly opposed. Finally, as we have seen, it is even argued, in the twentieth century, that reason has no place in science!

My claim is that the first kind of decline is all to the good, and should have been welcomed, rather than reluctantly conceded, whereas the second kind of decline is a disaster. Furthermore, what has been responsible, in the end, for the second kind of decline in the scope of reason, has been the misguided resistance to the first kind of decline in the power or authority of reason.[11] As long as reason is conceived of in authoritarian or oracular terms as a set of rules which deliver indubitable, unchallengeable decisions to us, thus taking our power to reach our own decisions away from us, then it is entirely understandable that it should seem highly desirable (or inevitable) that the scope of reason should be severely restricted. It may be acceptable that logicians and mathematicians, and even perhaps scientists in the context of justification, should be in a sense deprived of the power to decide what to accept and reject, being obliged to comply with the

[11]Hayek, for example, has failed to understand this simple but crucial point. Hayek argues for the need to restrict the *scope* of reason; but his arguments only have any validity insofar as reason is presupposed to be *powerful*, capable of delivering authoritative decisions. Reject this authoritarian conception of reason, and Hayek's reasons for restricting the scope of reason collapse (Hayek, 1967, ch. 5).

dictates of reason. That this should happen in our personal lives is surely horrendous. We should become slaves to reason.[12]

What ought to have been clearly recognized and acknowledged long ago within the rationalist tradition is that the authoritarian conception of reason is a perversion of reason.[13] The whole point of reason is to help us to act and decide as we really do want to act and decide: it is to enhance *our own capacity* to act and decide as we really want, not to wrench the capacity from us or to reduce it to the one decision to obey henceforth the dictates of reason. This is true in life; it is true in science; and it is even true in mathematics, and in logic. Confronted by a proposition P, which he cannot prove or disprove, the mathematician does not know whether he really wants to accept P or reject it. What a proof or disproof does is to reduce the big, uncertain decision to accept (or reject) P to a number of small decisions the mathematician knows he really does want to make (steps of the proof). A proof thus enhances the mathematician's power to make decisions that he really does want to make. It is of course obvious that mathematical thinking or problem solving – and especially the best, the most creative and original – cannot proceed in a way that is determined by explicitly stated rules. The most that could be said is that in

---

[12]An argument along these lines, designed to show that the more rational, the more rigorous we become, so the greater our loss of freedom, is expounded by O'Connor (1973, pp. 44–6).

[13]Demolishing authoritarian conceptions of reason (and of science) as irrational and non-humanitarian is one of the central tasks and achievements of Popper's great works, (1959, 1969). See, in particular, Popper's introduction, 1963: 'On the Sources of Knowledge and of Ignorance'. The present book develops further this theme of Popper's work. Reason itself (almost by definition) is valuably applicable to all that we do: if 'reason' seems to lead us systematically astray, or leads to undesirable results such as enslavement, or a regimented, unspontaneous, uncreative, unimaginative, insensitive or unloving way of life, a despotic society, or a cruel political programme, then it is not reason which thus leads us astray, but some form of irrationality masquerading as reason. It is vital that rationalists be highly critical of any conception of reason – especially as it is applied in practice – just because deformations in the conception of reason embodied in a person's life, in an institution (such as science) or in a society, result in deformations in the life itself, in the institution, in the society. My criticisms of Popper are thus very much in the rationalist, and the Popperian, tradition. A minor point of criticism (alongside the major criticisms) is that an element of authoritarianism lingers on in Popper's conception of reason, of method. For Popper's method, ideally, determines for us, in a fallible way, the best choice. It chooses *for* us, as it were. The view developed here is that putting into practice the heuristic methods of reason enhances our capacity to choose as we really desire: it enhances desirable spontaneity, creativity, freedom, and does not reduce freedom to the one decision to proceed in accordance with the methods of reason.

mathematics precise rules determine what is to count as a rigorous *proof*. But it must also be said that mathematicians actually become *irrational* if they merely slavishly obey the rules. For rigour requires that such rules be constantly criticized and revised when found wanting. And in fact one finds that as mathematics has developed, standards of rigour have been again and again revised and improved. Furthermore – and this point is entirely general – no slavish, uncritical obedience to any set of explicit rules, however good they may be, can conceivably amount to being rational simply because adoption of such rules can only be sensible and helpful in certain circumstances and for certain purposes. Slavish following of explicit rules is bound to lead us to follow the rules in contexts and for purposes beyond their domain of fruitful application, the outcome being a characteristic kind of irrational action (a point of great importance to be developed in the next chapter).

The fear that to become wholly rational in one's living and thinking is to become a slave to reason, losing one's freedom and one's soul, can perhaps best be put wholly to rest by the following consideration. The rules of reason (as reason is being conceived in this chapter), in order to satisfy the requirements of being relatively few in number and completely general, must be high-level *meta*-rules, which cannot tell us precisely what to do but can only tell us the sort of things we might try to do. Reason presupposes that we can already successfully put into practice a multitude of extremely diverse mostly implicit rules, methods, strategies in performing all the diverse actions we do perform in life – moving about in the world, perceiving, talking, writing and reading, cooking, earning a living, bringing up children and so on. All that the generally applicable meta-rules of reason can accomplish is to indicate – in a way that is open to infinitely many different specific interpretations – how almost infinitely diverse particular rules or methods already being put into practice can best be coordinated or marshalled so as to give us the best chances of solving our problems as we really want, thus achieving what is of real value to us. All of our thinking, feeling, desiring and doing inevitably proceeds in accordance with (mostly implicit) rules or methods of one kind or another whether we acknowledge this or not. This cannot be avoided. Acting spontaneously, or deciding to act spontaneously, does not alter this. The difference between living rationally and irrationally is thus not at all the difference between living in accordance with rules and in violation of rules. Rather, the difference is between exploiting strategies which offer us the best general help with realizing what is of value to us in life and failing to exploit such strategies (or systematically violating such strategies). And finally, it must be remembered that the rules of reason formulated above all tell us what to *attempt,* and do not necessarily specify an invariably performable action. There can thus be no question of slavishly obeying the above rules.

As long as reason is conceived of in quasi-oracular terms as, ideally, a set of methods which can be mechanically applied, scientific discovery, and creativity in general – inherently non-mechanical – are bound to seem beyond the scope of reason, as Reichenbach and Popper have both claimed. But the moment reason is conceived of in terms of non-mechanical methods which, if put into practice, give us the best chance of success (but which do not mechanically reach decisions for us), rational discovery and creativity become possible. In fact, not only do rational methods for creating good possible solutions to problems exist in science and in life (as the methods of rational problem-solving spelled out in this chapter make clear): methods of discovery, and methods of assessment, amount to two equally important, interdependent aspects of reason. Without good, explicit or implicit, methods of discovery, we will have no possible solutions to our problem to assess: without good methods of assessment, we will fail to choose well between good possible solutions.

The point can be put even more strongly: any conception of rationality which restricts itself to methods of assessment must be defective or irrational. A conception of reason can be said to be defective or irrational (as opposed to merely incomplete) if it can be shown to lead us *systematically* astray. The crucial point is now this. Granted that we tackle a problem P, defined in terms of an aim A, and methods M for assessing possible solutions (which specify what is to count as a solution to P), then, in general, it is to be expected that as we proceed our understanding of P will improve, our aim A will improve, and so too our methods M, our idea as to what is to count as a solution. As long as we adopt the philosophy-of-wisdom idea that (heuristic) methods of discovery H are fundamental to reason, this process of *improving* problems, aims and methods of assessment as we proceed can be quite naturally incorporated into the basic conception of reason: it is implicit in rules (1) and (2). Identifying rules of reason solely with any set of methods M for the assessment of possible solutions cannot allow for this necessary *modification* of such methods: and thus is to be rejected as irrational.

It is worth noting that even if the philosophy-of-knowledge view of science was correct, science having the fixed aim of improving knowledge of truth as such, and having fixed methods of assessment M, nevertheless a strong case could be made out for holding that scientific rationality cannot adequately be conceived in terms of methods of assessment alone. For, as Feyerabend has argued in a classic essay (1965), in order to assess a given theory well empirically, we actually need good rival theories to indicate severe tests. Thus empirical assessment of the given theory (in terms of methods M) actually requires good methods H of discovery to generate good rival theories to indicate severe tests!

The case becomes overwhelming once it is recognized that the philosophy-of-knowledge conception of science is unacceptable. The aim of science is not to discover truth *per se*, but rather *explanatory* truth, and more generally *valuable* truth. It is to be expected that as we proceed, our aims and our methods M of appraisal will improve. In order to be rational in pursuing science we must allow heuristic methods of discovery H, influencing our ideas concerning what our basic scientific aims ought to be, to modify our methods M of assessment. Any attempt to characterize the rationality of science solely in terms of fixed methods M of assessment must be defective or irrational. It must miss the essential thing: the way in which improving scientific knowledge enables us to improve our knowledge about how to improve knowledge. Both in science and in life, our basic aims are profoundly problematic: rationality requires that we improve aims and methods as we proceed (there being constant interplay between methods of discovery H and methods of assessment M).

I judge this point to be of such great and general importance that it is developed in the next chapter into a general 'aim-oriented' conception of reason.

CHAPTER FIVE

Aim-oriented Rationalism

The philosophy of wisdom can be formulated in a somewhat more general, and perhaps more adequate way by appealing to a more general notion of *aim-oriented rationality*. Instead of characterizing reason in terms of rules which help us to solve problems we can, more generally, characterize reason in terms of rules of action which, when put into practice, give us, other things being equal, our best hope of achieving what is genuinely desirable and of value. All problem-solving is aim-pursuing, but not all aim-pursuing is (conscious) problem-solving, since there is much that we do – even of value – that we do effortlessly, instinctively. A problem is a failed action, perhaps a deliberately over-ambitious action, an attempt to do in some new domain what has already proved to be successful elsewhere. (More specifically, any problem can be construed to be an aim A, a provisional route R to realization of A – the initial deed designed to realize A – and a barrier B which blocks the attainment of A along R. All real life problems – as opposed to problems set in exam papers – come with initial inadequate solutions, which may only need to be changed somewhat in order to become adequate.) Problem solving, then, is a special case of aim-pursuing. It is this which enables us usefully to modify our earlier notion of rational problem-solving to form a more general notion of *aim-oriented rational action*. Aim-oriented rationalism has the added advantage that it brings out much more explicitly and generally the important points concerning rationality made towards the end of the last chapter.

The basic idea of aim-oriented rationalism is extremely simple. It can be put like this. Whatever we are doing, our aims are quite likely to be more or less problematic. Contrary to what we may suppose, aims we are striving to realize may not be realizable, or may not be desirable (or may not be as realizable or desirable as somewhat modified aims we might pursue). Thus, whatever we are doing, in order to act rationally we must be able and ready, as the need arises, to improve our aims and methods as we act. Any conception of rationality which does not include this requirement concerning the need to *improve* aims and methods as we act must systematically lead us astray, fail to help us realize what is of most value to us (on all those occasions when we pursue unrealizable or undesirable aims). All such conceptions of rationality must thus be rejected. Quite generally, in order to be rational, we must be ready to look critically and imaginatively at our aims, to give ourselves the opportunity to discover how to pursue more desirable or more realizable aims; we must be ready to ask *why* we are pursuing the aims we are pursuing – in both the rationalistic, and historical or causal senses of 'why' – so that we may discover ways in which our aims can be improved; above all we must do all that we can to ensure that we are not *misrepresenting* to ourselves what aims we are pursuing – since if we *misrepresent* our aims to ourselves, our capacity to realize our actual aims rationally and successfully is seriously undermined. All this applies to whoever or whatever is doing the aim-pursuing, whether it be an individual person, a group of people, or an institution or social organization.

Any action that we perform – whether it be the action of an individual, of a group of people, or of an institution – has an aim and exemplifies a methodology. Thus, in improving our (personal and social) aims and methods of living we are improving what we do and are, our personal and social lives. It is of course always possible that a new, or an improved, method may first be discovered as a particular, successful new action – the more general method implicit in this action only subsequently being exploited by a variety of analogous actions. Even when a common set of methods inform a range of our actions, these methods mostly in practice remain implicit, it being perhaps impossible for us to formulate explicitly methods we successfully put into practice in doing such things as teaching, playing the violin, moving our limbs, constructing grammatical and meaningful sentences (all of which we must learn).

The prescription 'endeavour to improve your aims and methods as you live' is important just because it is simple, universal in its application, fundamental, and widely neglected.

In an attempt to demonstrate just how important, and how much neglected, how poorly understood, this simple idea of aim-oriented rationality is, I propose now to apply it to one important human endeavour, one important institutional enterprise, namely *science* – and more generally *academic inquiry*. What I propose to do is to begin with science as conceived of by standard empiricism and the philosophy of knowledge and show how *four successive applications* of aim-oriented rationalism transform standard empiricist science into philosophy-of-wisdom inquiry – into a version of this kind of inquiry, indeed, that constitutes a clarification and improvement of the version formulated in the last chapter. My claim is that the outcome of each of these four applications of aim-oriented rationality is a kind of science, a kind of inquiry, that is both of greater rationality and of greater human value – the end product thus being very much more rational and valuable than what we begin with. Each of these conceptions of inquiry – represented in figures 5b to 5e – upholds a different basic, overall intellectual *aim* for inquiry, and thus different basic, overall *methods*. Each of these conceptions of inquiry thus upholds a different conception of *intellectual progress*, and a different conception of the nature of the *problems* that rational inquiry is concerned to solve. Each conception of inquiry is more intellectually rigorous, more rational, than its predecessor, in the straightforward sense that it explicitly articulates, criticizes, and thus seeks to improve by rational means, assumptions that are substantial, influential and problematic but unacknowledged and thus only implicit in the conception of inquiry that is its predecessor. In other words, each step in the argument acknowledges more honestly what the aims of inquiry actually are, and what they ought to be: at the same time this involves acknowledging, what before was suppressed, namely the profoundly problematic character of the more honestly represented basic aims of inquiry. Acknowledging explicitly the *problematic* character of the actual aims of inquiry is important because, as a result, the problems associated with these aims can be explicitly discussed as an integral part of inquiry itself, this in turn holding out the hope that improved solutions to the problems can be rationally developed, the outcome being progressive improvement of aims and methods of inquiry as inquiry proceeds (the essence of aim-oriented rationalism).

Here, then, is the argument, set out in four steps, (1) to (4), each step exhibiting the same pattern of argumentation.

We begin with natural science conceived as having the basic intellectual *aim*, in the context of verification, of improving knowledge about the world, no presuppositions being made about the world, the basic *method* being to assess empirically testable conjectures about the world entirely impartially with respect to their empirical success and failure alone (see figure 5a). There is a fixed aim and a fixed method; there are essentially just two

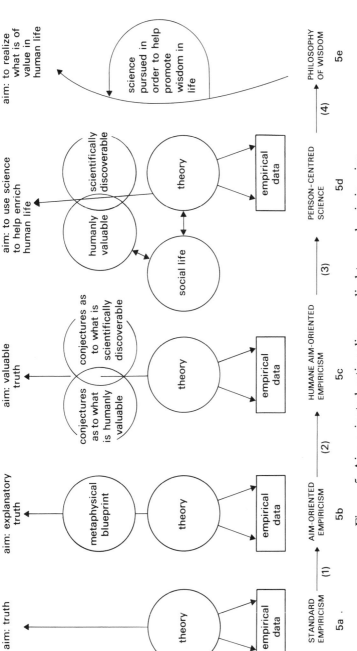

Figure 5 Aim-oriented rationalism applied to academic inquiry

domains of discussion, namely that of (a) observational and experimental results and (b) theory; an idea, in order to enter the intellectual domain of discussion of science, must at least be an empirically testable claim to knowledge (observational or theoretical).

1   This standard empiricist view seriously *misrepresents* the true intellectual aim of science. The aim of science is not merely to discover truth *per se*, nothing being presupposed about the nature of the truth to be discovered. A basic aim of science is to improve our *understanding* of the world. Science seeks *explanatory truth*. Even in the context of verification, the whole enterprise of natural science must presuppose (conjecturally, no proof or experimental verification of this being possible) that the universe is comprehensible to us, in some way or other. More specifically, modern natural science presupposes that there exists some kind of unified pattern running through all natural phenomena, it being a basic *aim* of physics to articulate this pattern as a testable, comprehensive, unified theory. In the absence of some such presupposition it is impossible to choose between the potentially infinite number of rival possible theories, equally acceptable from an exclusively empirical standpoint, that can always readily be formulated. In practice in science this situation is avoided by giving preference to those few theories that simplify and unify – so much so that empirically successful systems of propositions which do not simplify and unify do not count as 'theories' at all. This in practice commits science to the presupposition that unity exists in Nature to be discovered – a crucial point that will be established in chapter 9.

The aim of discovering some kind of unified, comprehensible pattern in the world, in terms of which phenomena can be explained and understood, is however profoundly *problematic*. Precisely what unified pattern does exist? In broad outline, what kind of pattern exists? What does it mean to assert that some kind of unified comprehensible pattern exists in Nature? Why should any such pattern implicit in the physical universe be comprehensible to us? Why should the universe be comprehensible at all? In terms of what concepts, what invariance and symmetry principles, is any such unified pattern to be characterized? What modifications need to be made to existing fundamental physical theories, and fundamental physical concepts (having to do with such things as space, time, energy, force, particle, field) in order that a clearer picture may be given of the conjectured unified pattern than at present? What grounds can we have for holding conjecturally that some kind of unified pattern is implicit in all phenomena? How can it be rational to commit science to such an article of faith, such a 'miracle-creed'? If some kind of unified pattern is implicit in all phenomena, more or less like patterns postulated by the fundamental theories of physics, how is it possible for there to be consciousness in the world? How is it possible for us to exist, experiencing, feeling, enjoying and suffering beings? How can there be colours, sounds, smells as experienced by us? If all that we are and do conforms to a fixed pattern of physical law, how can there be any free will? How can we be responsible for any of our actions, our thoughts, desires and decisions? How can there be purposiveness in the world? How can there be life? How can our lives have any meaning or value?

If science is to pursue its basic intellectual aim of improving our understanding of the world in a rational way – in a way which gives us the best hope in general of making progress towards the realization of this aim – then it is essential that these *problems*, associated with this aim, receive explicit rational discussion as an integral part of science itself. The intellectual domain of science must include *three* interrelated departments of discussion: (a) discussion of observational and experimental results; (b) discussion of testable theories; and (c) discussion of problems associated with the basic aim of improving understanding. This third department of discussion must seek (i) to improve the articulation of the problems just indicated and (ii) to propose and criticize possible and actual solutions, in an attempt to improve the basic *aim* of science. Untestable, metaphysical ideas need, in other words, to be proposed and criticized (in the light, of course, of current scientific empirical results and theories) from the standpoint of their capacity to improve solutions to problems that arise in connection with the search for scientific understanding, as an integral part of science itself, in an attempt progressively to improve the overall aim of science actually being pursued. And as the overall aim of science is improved, so too the methods of science can be improved. In short, as we improve our scientific knowledge and understanding of the world, we improve our (conjectural) knowledge and understanding of the domain of our ignorance; this enables us to improve the aims and methods of science; we thus improve our knowledge about how to improve knowledge – a vital feature of scientific method which helps to account for the relatively recent explosive growth of scientific knowledge. The philosophy of science (the enterprise of articulating aims and methods of science) thus turns out to be a vital part of science itself, which must evolve as an integral part of the evolution of scientific knowledge, if science is to be rational. (Standard empiricist philosophy of science, seeking to understand the rationality of science, but pursued as a discipline distinct from science, actually helps thereby to undermine the very thing it seeks to understand.)

As a result of correcting a serious misrepresentation of the basic aim of science (this in itself an application of aim-oriented rationality) scientific inquiry is revealed to exemplify, in a striking way, the basic idea of aim-oriented rationality (as briefly characterized above). I shall call this conception of science *aim-oriented empiricism* (see figure 5b).

According to aim-oriented empiricism, untestable ideas about how the world is comprehensible inevitably exercise a profound influence over what theories are accepted in science, and what research aims are pursued. Thus, if science is to proceed rationally, it is essential that such untestable ideas be articulated and criticized as an integral part of science itself, within the intellectual domain of science. Standard empiricism, on the other hand, demands precisely the opposite. Untestable ideas must be excluded from the intellectual domain of science. Scientists are, of course, permitted to propound and criticize untestable conjectures in the domain of discovery, to themselves and to each other, unofficially over coffee, as it were. What standard empiricism does not permit is the publication and criticism of untestable proposed solutions to problems inherent in the aim to understand in the official scientific literature, in the context of verification. This prohibition arises from a misguided attempt to preserve the rationality of science (misguided as a result of the basic aim of science being misrepresented to be discovery of truth *per se*). Actually the prohibition serves only to undermine rationality. Once the intellectual aim of science is acknowledged to be to improve understanding, it is clear that in order to pursue this aim *rationally* it is essential that we explicitly propose and criticize rival possible solutions to problems inherent in this aim, within the intellectual domain of science, in an attempt to improve the aim as we proceed. It is just this which we cannot do if we adopt the misconceived intellectual standards of standard empiricism.

Inevitably, if the scientific community pursues science in accordance with standard empiricism, scientific progress towards improved understanding of the world must tend to suffer as a result of the characteristic irrationality of standard empiricism.

The scientific community as a whole will fail to improve the aim of understanding the world, in an explicit, cooperative way, as a result of the failure explicitly to articulate and criticize diverse possible solutions to the problems inherent in this aim. A few individual scientists (such as Darwin, Faraday or Einstein) may individually improve their research aims in this way, and as a result they may well make many significant (testable) discoveries: they will be unable, however, to communicate to their fellow scientists how they have made such discoveries (since standard empiricism prohibits such communication). Many scientists will be brain-washed by standard empiricism and hence will fail to discover anything of much significance from the standpoint of improving our understanding of the world. As a result of the failure of the scientific community to articulate and criticize diverse possible solutions to the problems inherent in the aim to improve our understanding of the world, it is quite likely that the scientific community will accept, as a body, in a dogmatic and uncritical way, some set of answers to such problems, for a time science almost being defined in terms of these dogmatic answers, so that all research and all theorizing proceeds within the framework of these answers. Eventually *empirical* problems – clearly recognized by standard empiricism – may become so overwhelming, that a new, empirically more successful, comprehensive theory may be developed, violating the old solutions to problems of understanding but being in accordance with some new implicit set of possible solutions. It will not be possible to discuss the transition from the old to the new theory *rationally,* just because problems of understanding cannot be explicitly discussed within science. The transition will thus be made in the irrational way so brilliantly depicted by Kuhn (1962). Scientists, and historians and philosophers of science may even hold, with Kuhn (1970), that this is the way science ought to develop; it will not be seen as the unfortunate consequence of the irrational suppression of sustained imaginative and critical discussion of problems of scientific understanding, within the intellectual domain of science itself.

Failure to articulate the scientific aim of improving understanding may well lead science to degenerate into nothing more than the enterprise of predicting more and more phenomena more and more accurately. Those few scientists who prize the search for understanding above all else, and who protest, will tend to be dismissed as unscientific metaphysicians or philosophers. The united, cooperative endeavour to improve understanding is likely to disintegrate into fragmentary, disorganized, specialized research endeavours, with aims unrelated to each other, and often obviously defective, even though the specialized scientists who pursue these aims will not realize this. The vital task of attempting to interconnect these diverse, disorganized research aims will fall into disrepute as philosophical and unscientific. The scientific aim of improving our understanding of the world may even itself fall into disrepute. It may be declared unscientific. The world may be judged to be incomprehensible. An intellectual disaster will have overtaken science (from the standpoint of improving understanding) and most scientists will not even notice as long as much specialized knowledge, however trivial, is being accumulated.

The outcome of rejecting standard empiricism and adopting instead the more intellectually honest philosophy of science of aim-oriented empiricism might be, at first, for many scientists, disconcerting in that suddenly a wide range of intellectual defects of science leap to the eye that were before invisible. The eventual outcome would be, however, the (gradual) transformation of standard-empiricist science into something more closely resembling the natural philosophy of the seventeenth century.

2 Given the profoundly problematic aim for science of seeking to improve knowledge of explanatory truth, of seeking to improve our understanding of the world, we next need to ask, according to aim-oriented rationality: why are we seeking to realize this aim? What more general or more fundamental intellectual aim do we, and ought we to, seek to realize by its means? There is, I submit, an obvious general answer to this question. We seek to improve our knowledge of explanatory truth because, more generally and fundamentally, in doing science we seek to improve knowledge of *humanly valuable truth*, of value culturally or practically, explanatory truth being one kind of valuable truth. In pursuing science we seek to discover knowledge that is of the greatest value to humanity, of greatest value from the standpoint of developing a healthier, richer, more just, more civilized world. Above all our concern should be, so we may hold, to develop knowledge that is of most value to those whose needs are greatest – the poor, the ill, the suffering. But in any case, quite properly, scientific progress is assessed in terms of the extent to which knowledge of valuable truth is increasing, growth of knowledge of trivial truth only, however extensive, being not progress but rather stagnation and decadence. Thus, in assessing a potential contribution to science (in order to decide whether it deserves to be published in a scientific journal for example), value and truth factors must both

be taken into account. A contribution almost certainly true (and thus representing knowledge) may legitimately be rejected on the grounds of its triviality. A contribution almost certainly false may be accepted for publication (and even accepted as a great contribution to science) because of its potential value, its potential fruitfulness perhaps (even though false), or its value if by chance true. Thus considerations of value and truth cannot, and ought not to be, dissociated from one another even in the assessment of scientific results, and certainly not in the assessment of aims for research.

The aim of improving knowledge of *humanly valuable truth* is, if anything, even more profoundly problematic than the aim of improving knowledge of explanatory truth, the aim of improving understanding. What is of value? Whose values, whose needs and desires, ought to be given priority? What is there potentially of value in the domain of our ignorance, awaiting discovery by us, capable of being discovered and exploited in desirable ways by means of present methods? What will be of value to humanity in ten, fifty, one hundred years time? How can science do justice to the value of acquiring knowledge and developing technology, including medicine, that is most needed by the world's poorest people, and at the same time do justice to the value of improving knowledge and understanding of the universe for its own sake – especially as such knowledge is often esoteric, remote from the concerns of most people, and unlikely to lead to the development of new technology of any kind, let alone of the kind most urgently needed, as in the cases of astronomy, cosmology, high-energy physics? How can science contrive to give priority to the needs of the world's poor when scientific research is mostly financed by, and thus presumably responsive to the interests of, the world's wealthy? Does science do as government, industry or popular opinion bids: or does it seek to acquire knowledge of truth deemed by the scientific community itself to be of value? What research aims and priorities are to be taken up, who is to decide, and how (in terms of what criteria)?

The prescription of aim-oriented rationalism is essentially just the same as the already discussed in (1) above. If science and technology are to pursue their basic intellectual aim of improving knowledge of humanly valuable truth in a rational way then it is essential that the problems associated with this aim, receive explicit rational discussion as an integral part of science itself. The third department of the objective intellectual domain of science, recognized by aim-oriented empiricism, devoted to the discussion of problems associated with the aim to understand, will need to be broadened to include discussion of problems associated with the aim to improve knowledge of *valuable truth*. In order to choose aims rationally it is essential that we bring together discussion of factual (but possibly untestable) conjectures as to what exists to be discovered, and evaluative conjectures as to what it is genuinely desirable to try to discover, in this kind of way. And, furthermore, since the scientific community cannot claim to possess any special expertise which enables it to determine what is of human value better than the rest of us, the non-scientific community must be encouraged to take part in the sustained imaginative and critical discussion of aims for science within the intellectual domain of science. Whoever makes the actual decisions as to what research aims are to be pursued – individual scientists, heads of research laboratories and university departments, or grant-giving bodies – these decisions need to be informed and critically assessed by open discussion in scientific literature, and elsewhere. As a result of imaginatively and critically discussing the profound problems associated with the aim to improve knowledge of valuable truth, in an open, cooperative way, as an integral part of science – and as a result of actively promoting, and responding to, such discussion in society – the scientific community may be able, as it proceeds, gradually to improve its actual overall aims and methods, its priorities, from the standpoint of discovering truth of most value to humanity. This view of science may be called *humane aim-oriented empiricism* (see figure 5c).

Standard empiricism excludes discussion of values from the intellectual domain of science in an attempt to preserve scientific rationality. Actually this serves only to undermine rationality, in that it places influential and problematic ideas (concerning what is of value) beyond criticism within science.

Inevitably, if the scientific community does proceed in accordance with standard empiricism, scientific progress towards acquisition of knowledge of most value to humanity must tend to suffer in characteristic ways as a result of the characteristic irrationality of standard empiricism. The scientific community, together with the non-scientific community, will fail progressively to improve the aim to discover truth of most value to humanity in an open, explicit, cooperative way, as a result of the failure to articulate and criticize diverse possible solutions to the problems inherent in this aim. A few individual scientists may seek to improve their individual research aims in this way. Unfortunately, as they do so, they are increasingly likely to fail to get funds to make such research possible. Furthermore, they will be unable to communicate to their fellow scientists the need for the scientific community as a whole to improve the basic intellectual aim of science in this way, since standard empiricism prohibits such communication. Indeed, the attempt of such scientists to convince their fellow scientists of the need to pursue science more rigorously, by articulating and criticizing possible solutions to the problems inherent in the aim of discovering *valuable* truth will be vehemently opposed by the majority who, accepting standard empiricism, will see the intrusion of moral and political ideas and problems into science as a threat to the objectivity, the rationality, the intellectual integrity of science. In seeking to preserve the intellectual integrity of science, with the best of intentions, they will be preserving a characteristic kind of irrationality in science. The failure of the scientific community as a whole actively to promote open, imaginative and critical discussion of problems associated with the aim of discovering *valuable* truth will almost inevitably result in that community pursuing research aims and priorities that merely reflect the interests and the values of those sufficiently rich and powerful to pay for scientific research – industrial concerns and governments of wealthy nations. The scientific community must tend to fail to develop and pursue research priorities that reflect the interests of the world's poor, hungry and suffering; general adoption of standard empiricism will shield the scientific community from an awareness of the extent of their betrayal of humanity. Indeed, the scientific community may well, with a clear conscience, and with a full sense of scientific righteousness, pursue goals that are only of much value to the scientific community itself, because the truth discovered happens to be of interest to some scientists (even though otherwise relatively uninteresting and useless) or of value because of the gladiatorial interests inherent in science: Nobel prizes are won, scientific reputations are made, careers are advanced. In choosing research projects, scientists may well be influenced more by their concern to advance their position in the pecking order of international scientific reputations than by their concern to help alleviate the harsh conditions of life experienced by millions in central and south America, in Africa and in Asia, by relevant technological, medical or agricultural research. Furthermore, the scientific community must tend to fail to organize the awarding of scientific honour and status in such a way that these are to be achieved by successful research of maximum benefit to humanity – even though not necessarily scientifically fashionable or glamorous.

Occasionally, dramatic improvements in scientific understanding
may also lead to dramatic advances in technology of just the kind
most needed by those whose plight is the greatest. But as scientific
understanding has improved, the likelihood of scientific discoveries being of value in both these ways becomes progressively
less and less. A major problem arises: how to balance the value of
improving understanding against the value of relieving suffering.
Standard empiricism, obscuring the need to give sustained
intellectual attention to this agonizing problem, must tend to
produce a science that merely predicts more and more phenomena
more and more accurately, thus failing both to improve understanding and to lead to the development of the kind of technology
most urgently needed. Scientific and technological research will
come to suffer from all the defects discussed in chapter 3.

One point must be emphasized. I am not arguing that these
failings of science arise because scientists are wicked, cowardly or
selfish. And nor am I arguing the other side of the coin to this: that
if science is to avoid these failings scientists need to be nobler,
more compassionate and courageous. The argument is entirely
different. It amounts to this. The aim of discovering truth of most
value to humanity is inherently profoundly problematic. If science
is to make good progress towards achieving this aim, then diverse
conjectures about how to solve the problems of the aim must be
persistently articulated and criticized within the intellectual
domain of science, every attempt being made to interconnect this
discussion conerning aims with research aims actually being
pursued. The scientific community does not only need to learn at
the level of empirical data and theory: in addition it needs to learn
at the level of aims, by means of the standard procedure of
conjecture and criticism. If the scientific community excluded
from the intellectual domain of science the open proposing and
criticizing of theories, decisions about what theories science is to
accept being reached by committees, it would occasion no surprise
if, as a result, science manifested a profoundly defective theoretical
knowledge. Likewise, if the scientific community excludes from
the intellectual domain of science the open proposing and
criticizing of conjectures about aims, decisions about what
research aims are to be pursued being reached by (grant-giving)
committees, it should occasion no surprise if, as a result, science
manifests a profoundly defective choice of aims. Thus any
scientific community which adopts standard empiricism, and which
thus excludes discussion of aims from the intellectual domain of
science, must inevitably come to pursue profoundly defective

aims, even though the community is made up of reasonably 'noble' individuals. If standard empiricism is indeed built into the institutional structure of modern science, then it should occasion no surprise whatsoever if modern science also exhibits intellectual and humanitarian failings of the kind just described. In this case what needs to be done is to remould the intellectual/institutional structure of science in accordance with *humane aim-oriented empiricism*. In this way we may be able to institutionalize an intellectually and morally nobler, more compassionate and courageous science, without it being necessary to make unrealistic assumptions about scientists.

3   Given the (problematic) aim for science of seeking to improve knowledge of *valuable truth*, we next need to ask – according to aim-oriented rationality: why are we seeking this aim? Why ought we to be seeking this aim? What more general or more fundamental aim do we seek to realize by its means? Once again, in general terms the answer is surely obvious. Science seeks to improve knowledge of valuable truth in order to make it available to people so that it may be used by them in order to help enrich the quality of their lives. Knowledge and understanding, however potentially significant or important, mean nothing as long as they remain in scientific journals not used or appreciated by anyone. This is obviously the case as far as technological discoveries are concerned: but it is also true of contributions to 'pure' science. What ultimately matters, from this latter standpoint, is the curiosity, wonder, knowledge, and understanding achieved by, and shared between, people. Pure science is of value insofar as it *is* this, or contributes to this.

As before, however, this aim of science is profoundly – and notoriously – problematic. Science has unquestionably been successfully used by millions of people to enrich their lives. Scientific and technological discoveries have made it possible to create the industrially-advanced modern world, to be found in Europe, the USA and elsewhere, with all its amenities, freedoms, health and longevity – incomparably more wealthy a way of life than anything to be found in earlier times. And furthermore, the immense advances in scientific knowledge and understanding achieved during the last century or so enable each one of us to explore imaginatively this mysterious cosmos in which we find ourselves – thus enriching our life – to an extent far beyond what was possible in earlier times. The immense diversity of living forms on earth, their diverse character and ways of life; the miracle of the slow evolution of life on earth during some 650 million years at least; the far reaches of space, with its stars, pulsars, galaxies, quasars, black holes; the far reaches of time, stretching back even to the first few moments of the cosmos; the ultimate nature of matter, the strange domain of the quantum, and the unifying patterns of natural law embedded in all phenomena: these extraordinary and remote aspects of our world have been thrown open to our personal inspection by modern science.[1] But alongside these successes, there are the notorious (already discussed) problems and failures associated with the aim of using science and technology to enrich life – problems associated just as much with the cultural as the technological aspect of science. Thus non-scientists who are otherwise highly educated often profess ignorance of and hostility towards science, as C.P. Snow once reminded us (Snow, 1964). Most people in industrially-advanced countries, though surrounded by the products of science, are probably merely bemused by and somewhat resentful of the esoteric, autocratic mysteries of modern science, while most of those who live in the third world can have few opportunities to learn about science. It is by no means clear, in any case, how the scientific vision of the world can enrich our lives. According to this vision, so it would seem, everything – including ourselves – is made up of a few different sorts of fundamental particles interacting in accordance with precise law. Our freedom, our consciousness, our individuality, all the colour, richness, meaning and value of life, seem to fade away entirely, leaving nothing but leptons and quarks! It is not obvious that such a vision of the world is life-enhancing.

The prescription of aim-oriented rationalism is essentially just the same as that discussed in (1) and (2) above. In view of the profoundly *problematic* character of the aim of enriching life with science, it is essential, if science is to pursue this fundamental aim in an intellectually rigorous way, that science does all it can to promote and sustain explicit, imaginative and critical discussion of problems associated with the aim, both within science, and in the community as a whole – a conception of science that may be called *person-centred science* (see figure 5d).

As before, standard empiricism and the philosophy of knowledge, in excluding discussion of problems associated with the aim

---

[1]Excellent recent non-technical expositions of these topics are to be found in Attenborough (1981); Silk (1980); Weinberg (1977); Davies (1979); and Mulrey (1981);

to enrich human life from science, undermine both the rationality of science, and the capacity of science to enrich human life.

4 We come now to the crux of the entire argument. Given the (problematic) aim of exploiting science in order to enhance the quality of our lives, we once again need to ask – according to aim-oriented rationalism – precisely as before: why are we, and ought we to be, seeking to realize this aim? What more general or more fundamental aim do we seek to realize by its means? Once again, the answer, in general terms, is obvious. We endeavour to realize this aim because, more generally and more fundamentally, we endeavour to realize what is of value to us in life, as we live. The pursuit of science and technology is but an aspect of, a tributary to, our central and fundamental pursuit of value in life.

The vital point that now needs to be recognized is that this fundamental aim of realizing what is of value in life, is, if anything, even more inevitably and profoundly *problematic* than the previous aims for inquiry discussed in (1), (2) and (3) above. What ultimately is of value, given the brevity of life, given that all that we do, experience and suffer in the end comes to nothing, and given that the world really is more or less as modern science tells us it is? What is it in life that we should seek to attend to, to realize, to cherish and to love?

In a changing world that we only partly know and understand, inevitably our personal and social aims must have their problematic aspects. Quite generally, aims that we are pursuing may not be in our best interests because – despite appearances to the contrary – they are unrealizable in principle, unrealizable in practice, not as desirable or as realizable as modified or different aims available to us, not the most desirable or realizable means to more general or distant goals we seek to realize, undesirable because of unforeseen, undesirable consequences, undesirable as a result of being inadequate resolutions of conflicts between desires or aims.

Thus, quite generally, in order to pursue aims rationally, in such a way that we give ourselves the best chances of realizing what is really of value to us, it is essential that we acknowledge the inherently problematic character of our aims, and the possibility that we may have misrepresented to ourselves the problematic aims we are actually pursuing. It is essential that we imaginatively articulate and critically assess possible solutions to problems inherent in the aims we pursue, as an integral part of our aim-pursuing, in an attempt to improve our aims and methods as we proceed. In endeavouring to help us realize what is of value in life, the fundamental intellectual task of organized inquiry – of science, technology, scholarship and education – is to help us, individually and cooperatively, to improve our aims and methods in this way.

It is above all the humanities and the diverse branches of social inquiry that have this fundamental intellectual task of helping us improve our personal and social aims and methods in life. The social 'sciences' are not *sciences* at all. They are social *methodologies* or social *philosophies* – concerned to articulate, and to criticize diverse actual and possible aims and methods for our diverse personal and institutional endeavours. What (aim-oriented) *scientific methodology* is to science, so *economic methodology* is to actual *economic endeavour* in the real world, *political methodology* is to *politics, sociological methodology*, more generally, is to our diverse *institutions* and *social endeavours*. On this view, the *sociology of science* is precisely the the same thing as the *methodology of science* (or the philosophy of science). This book thus itself exemplifies philosophy-of-wisdom social inquiry. What I attempt here for the academic enterprise, other social inquirers need to attempt for other institutions – government, industry, the media, the law, international relations. Economics, political philosophy, sociology, psychology, history, anthropology, the study of international relations, philosophy, the study of industrial relations, education studies, the sociology or philosophy of science, of art, of literature, of drama, of religion: all are concerned to help us build cooperative aim-oriented rationality into our diverse personal and social endeavours, thus giving ourselves better opportunities to realize what is of value to us in life. Insofar as these diverse academic disciplines seek to acquire knowledge, this is acquired in order to further the fundamental intellectual task of helping us in practice improve aims and methods in life. All these diverse disciplines may be regarded as aspects of *utopian studies* – the enterprise of imaginatively articulating and severely criticizing possible and actual aims and methods for humanity as a whole.

On this view, then, those who hold that the methods of the social sciences are similar to those of the natural sciences (the pro-naturalists) and those who hold that they are different (the anti-naturalists) are both wrong. Social inquiry (not *science* at all) seeks to establish unity of method throughout all social endeavour, including science. It seeks to help enhance wisdom in life, by helping to build aim-oriented rationalism into the fabric of personal, institutional and social life.

The outcome of this entire argument (1) to (4) above, is thus rational inquiry as conceived of by the philosophy of wisdom (see figure 5e). As a result of four successive applications of aim-oriented rationalism, standard empiricist science has been transformed into philosophy-of-wisdom inquiry.[2] This new version of the philosophy of wisdom incorporates, and in some respects improves on, everything depicted in chapter 3. It clarifies what ought to be the aims and methods of natural science and of social inquiry. It clarifies how natural science and social inquiry ought to be related to one another, and to personal and social life. And the entire argument establishes, in outline, that this improved version of the philosophy of wisdom depicts a more intellectually rigorous and a more humanely valuable kind of inquiry than that depicted by the philosophy of knowledge.

It deserves to be noted that, just as at stages (1) and (3) above, so now at stage (4) of the argument, it can be shown that the attempt to pursue inquiry in accordance with standard empiricism and the philosophy of knowledge is damaging in both intellectual and human terms. This attempt leads economists, sociologists, psychologists and others, whether pro- or anti-naturalists, to pursue social inquiry as the academic enterprise of improving knowledge about diverse aspects of social phenomena, in a way that is more or less dissociated from society itself. The chief intellectual task becomes to solve sociological (or economic or psychological) problems of knowledge, not practical problems encountered by people in life. Thus the vital social and methodological task of social inquiry of helping us build aim-oriented rationality into our personal, institutional and social lives, demanding for its fulfilment active involvement with social life, is prohibited.

Among other advantages, aim-oriented rationality is more helpful than 'problem-solving' rationality when it comes to resolving conflicts between people. The way we formulate our problem depends on what we take our aim to be. Thus two people, caught up in some common enterprise, but with conflicting aims, will formulate their common problems in different ways. As a result, each may regard the other as illogical, merely self-interested, engaging in trickery, bluff, propaganda. This does not help cooperative rationality to develop. By contrast, putting aim-oriented rationality into practice enables us to avoid such unnecessary, destructive misunderstandings, and helps us – if we so wish – to develop gradually more cooperative ways of resolving our conflicts. In roughly increasing levels of desirability, conflicts

[2] A dramatization of this argument can be found in Maxwell (1976b).

between people are settled by: force, threat, manipulation, some more or less arbitrary procedure (such as tossing a coin or voting), bargaining, the cooperative discovery of the most desirable, just resolution. The general adoption of the aim-oriented conception of reason is in all our long-term interests in that it offers us the best hope of increasing our capacity to resolve our conflicts in rather more desirable ways – even though, of course, it provides no magic procedure for resolving conflicts.

Aim-oriented rationality can be regarded as a kind of empiricism, in that it specifies a general methodology for 'learning from experience'. However 'experience' must be understood here in commonsense terms as that which is acquired through action, doing things, living, actively engaging in some enterprise. And what is learnt is how to do things, how to live, how to achieve that which is desirable and of value, varieties of wisdom.

Looked at in this way, aim-oriented rationality transforms philosophy into an essentially empirical, practical enterprise. In actively engaging in some endeavour, in pursuing aims, and adopting methods in order to realize aims, we in effect put into practice a 'philosophy' of our endeavour, whether we are aware of this or not. Other things being equal, we give ourselves our best chances of learning from experience if we articulate, imaginatively develop, and scrutinize philosophies that we put into practice in our various endeavours in life. Above all, by developing a *tradition* of accurately articulating our actual philosophies, our actual aims and methods in life, we make it possible for us to learn from each other's experience, from each other's successes and failures. This essentially is what aim-oriented rationality amounts to: clearly it gives to 'philosophy' the practical task of helping us to improve our aims and methods as we live – of enhancing our capacity to learn from experience. In order to be successful, of course, it is essential that we are able to be honest, to ourselves and to each other, about what we are doing, what our actual aims and methods are, what it is we desire and feel. A basic task for inquiry, according to the philosophy of wisdom, is to help us to develop a society, a world, in which such honesty is encouraged to flourish as opposed to being penalized.

One feature of the above argument, (1) to (4), is that it repeatedly establishes that academic inquiry seriously misrepresents the basic intellectual aim of inquiry. The real but profoundly problematic aims for science of discovering *explanatory* truth, or *valuable* truth, are misrepresented by standard empiricism to be the apparently unproblematic aim of discovering truth *per se*.

In honour of Freud, any aim-pursuing endeavour, whether personal or institutional, that misrepresents its aims in this way may be said to be suffering from *rationalistic neurosis*. Thus, according to the above argument, theoretical physics, and science more generally, suffer from rationalistic neurosis, as depicted in figures 6a and 6b.

Quite generally, in order to act rationally, it is essential to be open to the possibility that one's aim-pursuing endeavours suffer from rationalistic neurosis. This is because rationalistic neurosis – misrepresentation of aims – is almost bound to occur, and once established, can be profoundly damaging.

It might be thought that the assertion that people and institutions tend to suffer from misrepresentation of aims, from rationalistic neurosis, must amount to a highly speculative, dubious psychoanalytic theory, Freudian or post-Freudian, of uncertain scientific standing. This entirely misses the point. Representing to oneself or others the goal one is pursuing is itself a goal-directed endeavour, which may succeed or fail like any other. Furthermore, it is often highly problematic to represent or characterize accurately the goals actually being pursued by animals, people, or institutions. We cannot declare merely that the goal is what results as the outcome of the being's actions – since this leaves open the question of precisely how any outcome is to be characterized, and in any case does not take into account the important possibility that the being may pursue some goal G and may fail to achieve it. In attributing a goal to any being, we invariably offer an interpretation of the being's actions; in many cases, it may be hard to choose between a number of such rival interpretations, rival attributions of goals to the beings in question. One principle can be employed in deciding what goal or goals a being is in fact pursuing: a being is to be interpreted as in fact pursuing that goal (or goals) which, if postulated, makes the best overall sense of what the being *does* (taking internal 'imaginings' etc., and misunderstandings into account), and which, at the same time, accords best with what the being has done in the past, and how it has come to be. It is something like this principle, implicitly understood, which leads us after Darwin, to attribute to animals the overall goals of survival and reproduction.

It is thus an uncertain, theoretical matter to characterize correctly goals in fact being pursued by animals, people and institutions, even when the person in question is oneself. Goals may be misrepresented for Freudian reasons, as a result of the repression of problematic aims (as in the case of science). Equally, goals may be misrepresented merely because the difficult task of representing goals accurately has not yet been accomplished. In holding that people misrepresent the goals they are pursuing, whether for Freudian or non-Freudian reasons, we do not need to believe in the existence of such Freudian entities as the id, the ego and the superego. In holding that institutions misrepresent goals they are pursuing, we do not need to imply, in any illegitimate sense, that there exist entities such as the institutional 'mind', 'unconscious' or 'id'. Even if we always sought with absolute intellectual integrity to represent accurately, to ourselves and others, the aims we pursue, it is to be expected that we will fail on occasions to represent goals we are actually pursuing correctly. We can make mistakes about what it is our actions are directed towards bringing about, just as we can make mistakes about any other factual matter about some aspect of the world. It behoves us, as rationalists, to suspect we have misrepresented our aims just as it behoves us to suspect that we have misrepresented other factual matters about the world: we need to adopt this sceptical attitude in order to make it possible for us to improve our representations of these factual matters.

Granted that we seek to live rationally, and develop rational institutions, it is especially important that we recognize that misrepresentations of goals are bound to happen, because of the tendency of misrepresentations of goals to sabotage reason – transforming it from something useful into something counter-productive. This will happen whenever aims are misrepresented in the characteristic manner of *rationalistic neurosis* as illustrated in figures 6a and 6b.

Rationalistic neurosis is damaging in a number of ways. The more 'rationally' the declared aim C is pursued – that is, the more thoroughgoing the corresponding methodology M_C is put into practice – the worse off the person or enterprise is, from the standpoint of realizing the genuinely desirable aim A. Even though B rather than C is pursued, B must nevertheless be pursued somewhat furtively and ineffectually in order to maintain the fiction that C and not B is being sought. Failure to acknowledge the actual aim B means that problems associated with B cannot be recognized, as a first step to their resolution, and the pursuit of A. Instead of recognizing and seeking to solve those problems that need to be solved if A is to be pursued rationally and realized, the person or enterprise, as a result misrepresenting

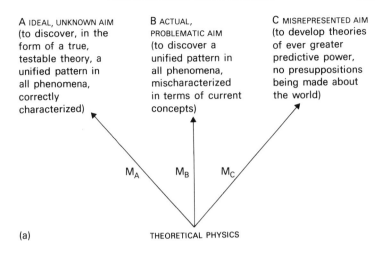

(a)

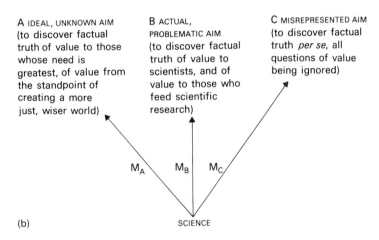

(b)

Figure 6(a) Rationalistic neurosis of theoretical physics

Figure 6(b) Rationalistic neurosis of science

the aim to be C, recognizes and tries to solve a number of what may be called *neurotic problems* – problems which, if solved, only make matters worse. There are all the neurotic problems associated with pursuing and realizing the declared goal C, defining and putting into practice the methodology M_C best designed to help realize C. There are the neurotic problems of explaining and understanding how real success – steps towards B or A – can be steps towards C, taken in accordance with M_C. In all these ways, reason is counterproductive. The more rationally the declared aim C is pursued, the more nearly problems associated with realizing the declared aim C are solved, so the worse off the person or enterprise is, from the standpoint of realizing what is really of value, namely A. As long as the rationalistic neurosis persists, the vital activity of articulating and attempting to improve aims and methods – the essence of reason – becomes counterproductive, or at best sterile.

This counterproductivity of reason is likely to have a further damaging consequence: the experience of reason being useless or even harmful may lead the person or enterprise suffering from rationalistic neurosis to hold that reason deserves to be ignored: and as a result it becomes very much less likely that the rationalistic neurosis will be detected and overthrown – since this does require reason, authentic reason. As a result, rationalistic neurosis, once established, is very likely to persist, just because it has the peculiar capacity to discredit the very tools that are needed to overcome it.

It deserves to be noted that even pure mathematics, traditionally held to exemplify reason, intellectual rigour, at its finest, actually suffers from severe, damaging rationalistic neurosis. As a result of being pursued within the context of the philosophy of knowledge, pure mathematics is traditionally interpreted to have the aim of increasing knowledge of mathematical truth. At once problems arise as to what mathematics can be knowledge *about*. Empiricism fails to explain how mathematical results can be *proved*. Platonism fails in this respect even more lamentably: since no one has ever observed Platonic mathematical entities, all knowledge of such entities ought – so one would suppose – to be irredeemably speculative, and not capable of being proved at all. Logicism and formalism fail even more dismally. Logicism holds mathematics to be merely ever more intricate elaborations of logical truisms: formalism holds mathematics to be devoid of meaning altogether. If either were true, mathematics would be the intellectually disreputable enterprise of discovering ever more intricate ways of asserting *nothing*.

These, and other related, traditional problems in the philosophy of mathematics are all *neurotic* problems arising from a misrepresentation of the aim of mathematics. Viewed from the perspective of the philosophy of wisdom, the idea that pure mathematics has the aim of acquiring knowledge of anything actual can be dismissed out of hand. Rather, pure mathematics has the aim of contributing to wisdom by developing, systematizing and unifying problem-solving methods applicable to as wide a range of important actual problems as possible. Mathematics explores significant, problematic possibilities, not anything actual at all. Mathematics exists because apparently very different real-life problems are solved by means of common or analogous methods: mathematics develops and unifies such common, widely applicable methods.

As a result of being pursued explicitly in accordance with the philosophy of wisdom rather than knowledge, mathematics would be transformed. Teaching, research, comprehensibility and availability of mathematics would be improved. Mathematics would be able to contribute far more effectively, more rationally, to the task of applying reason to the realization of what is of value in life.

The above theory of aim-oriented rationality and rationalistic neurosis provides us with a radical reinterpretation of psycho-analytic thought. Instead of regarding Freud, Adler, Jung and the diverse post-Freudian contributors to psychoanalytic thought as potential scientists, seeking to contribute to psychology, to our knowledge and understanding of the human psyche, we may instead regard them as *methodologists*, making contributions to the theory of aim-oriented rationality. A major claim of this book is that the social and humanistic disciplines quite generally – economics, political philosophy, the study of industrial relations, philosophy/sociology of education, of art, and of inquiry – need to be pursued and understood not as sciences at all, but rather as methodologies of our diverse social endeavours, helping us pursue these endeavours more rationally and successfully. Interpreting psychoanalytic theory along these lines is then just a special case of the general thesis.

Psychoanalytic theory interpreted in this aim-oriented ration-alistic way has several advantages over psychoanalytic theory interpreted as an empirical theory about human nature, an intended contribution to knowledge. It vastly enhances the scope of psychoanalytic ideas, in that such key notions as repression, rationalization and neurosis become applicable to any sufficiently sophisticated aim pursuing entities – in particular, as we have seen, to institutions – and not just to people. It vastly increases the acceptability, the epistemological status, of key psychoanalytic ideas. Instead of the general theory of neurosis being an empirical theory about human nature of dubious scientific standing, it becomes essential to rationality to suspect aim-pursuing endeavours of having succumbed to neurosis, this having happened even to science itself!

The proposed reinterpretation frees psychoanalytic theory quite naturally from some dubious claims of Freud – in particular his theory of the id, the ego and the superego, his theory of the death wish, and his view of the paramount importance of the sexual drive in human life (this probably having more to do with general hypocrisy about sexual matters in Freud's time rather than anything as basic as our biological nature, derived from evolution and our animal past). The proposed reinterpretation makes it much easier to see how psychoanalytic theory could quite naturally become a part of biology, Darwin's theory of evolution, and psychoneurology, all major hopes of Freud. It is reasonable to hold that a major factor in human evolution is the evolution of self-consciousness. A major part of self-consciousness may reasonably be held to be the representation to oneself of the pattern of goals one pursues in life – animals, by contrast, at most only representing to themselves goals actually being pursued. Our discussion above makes it quite clear that as such sophisticated representation of goals develops, so misrepresentation of goals is almost bound to develop as well. Neurosis may thus be maintained to be an almost inevitable teething problem of the early growth of self-consciousness – a point not without implications for psychoneurology. In order to make intelligent guesses as to how self-awareness and self-misawareness may have developed, within a framework of ethology and Darwinian theory, it becomes essential to attend not only to anthropological studies of hunting and gathering tribes, but also to such studies as those of Jane Goodall (1971) and her followers into the way of life of the chimpanzee – which may be taken to be close to pre-self-conscious human life. Such an approach to understanding human nature – via its biological and historical evolution – is bound to lead to a view of ourselves and our problems different from that of Freud. Observation of chimpanzees for example, may lead us to suspect that male competitiveness is as important a (misrepresented) basic drive in human affairs as sex. The aim-oriented rationalistic interpretation of psychoanalytic theory also has implications for psychotherapy: it implies that the basic task of therapy is to develop better strategies for living rather than to uncover the underlying 'causes' of the neurotic problems. Indeed, one consequence of putting the philosophy of wisdom into practice is so to change both therapy and education that the distinction between the two all but disappears.

Our discussion has revealed a situation rich in irony. Psycho-analytic theory has been criticized for failing to attain the high intellectual standards of science (Popper, 1963, ch. 1; Cioffi, 1970, pp. 471–99). It turns out that the thing is all the other way round. It is science that fails to attain the high intellectual standards of post-Freudian rationality, in that science suffers from rationalistic neurosis. On the other hand, it must be admitted that proponents of psychoanalytic theory have themselves misrepresented and misunderstood quite radically the goals and nature of their own discipline, in that they have construed it to be a branch of knowledge rather than a branch of aim-oriented rationality. The experts of misrepresentation have succeeded in seriously and damagingly misrepresenting the goals of their own discipline. Psychoanalytic theory itself suffers from rationalistic neurosis!

This aim-oriented rationalistic version of the philosophy of wisdom can be regarded as being derived from many sources: Lao Tzu, Socrates, Jesus of Nazareth, Einstein, Freud, Darwin, J.S. Mill, Kropotkin, Dewey, Fromm, Popper, amongst others. Some may regard Marx as an important source, especially in the light of that aspect of Marx's thought summed up in his eleventh thesis on Feuerbach: 'The philosophers have only *interpreted* the world, in various ways; the point, however, is to *change* it.' The two most important general sources are however: the Rationalist Enlighten-ment movement associated with such figures as Bayle, Voltaire, Diderot, Condorcet, Hume, Kant; and the Romantic movement, associated with such figures as Vico, Rousseau, Goethe, Beet-hoven, Blake, Wordsworth, Keats, van Gogh, William Morris, Tolstoy, and many contemporary and later writers, artists, musicians. An important achievement of the aim-oriented ration-alist version of the philosophy of wisdom is that it provides us with a synthesis, a unification of these two great, but conflicting, humanitarian movements. The Rationalism of the Enlightenment upheld versions of the philosophy of knowledge. It thus upheld associated intellectual ideals such as anti-authoritarianism; scepti-cism; belief in the value of reason; objectivity; method; logic and evidence; impersonal observation and experimentation; science and scholarship; the pursuit of impersonal, progressive factual knowledge. Rationalism tended to be suspicious of imagination, subjectivity, spontaneity, instinct and inspiration, personal experience, personal feelings and desires, passion. The Romantic movement rebelled against what was taken to be this Rationalist disparagement of much of value in human life. Romanticism thus sought to uphold and celebrate the value of that which Rationalism undervalued: imagination, inner experience, personal feelings and desires, spontaneity, instinct and inspiration, self-expression in art, in literature, in music, in love – and in all of life. It sought to celebrate the basic value of life as it is actually experienced and lived. Unfortunately, Romanticism made the mistake of assuming that Rationalism did indeed stand for what it claimed to stand for – namely genuine reason. As a result, in celebrating the central value of personal experience and life, Romanticism took itself to be celebrating the non-rational or even the irrational: it even on occasion *advocated* irrationalism. A disastrous split developed between the Rationalist movement – associated with science, technology, scholarship, the universities and much education, and the Romantic movement – associated with much literature, drama, art, music, and some education, psychotherapy, politics and religion. Both suffered as a result – both being more or less irrational and undesirable. What ought to have been realized long ago is that the Rationalist espousal of the philosophy of knowledge is actually irrational precisely because it excludes Romantic intellectual ideals of motivational and emotional honesty, truth to personal experience, imagination employed in the exploration of possibilities of value. The philosophy of wisdom is intellectually more rigorous than the philosophy of knowledge precisely because it incorporates such vital Romantic intellectual values. Aim-oriented rationalism heals the traditional split between Rationalism and Romanticism – the split between Snow's two cultures. It puts the two together, very much improving each as a result, the two uniting to form a coherent intellectual-cultural movement (Rational Romanticism or Romantic Rationalism), capable of devoting itself far more effectively to the cooperative realization of value in life.

If any one person in history deserves to be credited with discovering, practising and advocating the philosophy of wisdom, that person is Socrates. The following eight points about Socrates' thought and life constitute, in particular, especially striking grounds for making this claim. (1) Socrates' basic problem was this. What is the good life? How ought we to live? What is genuinely of value in life and how is it to be achieved? (2) For Socrates, this basic problem of living was more fundamental and important that the problems of cosmology or natural philosophy –

the standard problems of the Presocratic philosophers. (3) Basic to Socrates' thought and life is his claim that we are all more or less ignorant as to how to live well, he, Socrates, in particular, sharing in this common ignorance. As Socrates tells us in Plato's *Apology* – generally presumed to be reasonably accurate historically – Chaerephon, an impetuous friend of his, had the audacity to ask the oracle at Delphi whether anyone was wiser than Socrates. The oracle replied that no man was wiser. After pondering this judgement for some time, and after questioning others who claimed to possess wisdom, in an attempt to refute the oracle, Socrates came to the conclusion that what the oracle meant was that he, Socrates, was wiser than others in that he at least knew full well that he lacked wisdom. Socrates decided to become, as he put it, a gadfly, stinging his fellow Athenians into recognizing the inadequacy of their claims to wisdom, so that they might at least possess the wisdom of acknowledged ignorance. In other words, Socrates' central discovery and insight is just the fundamental presupposition of the philosophy of wisdom. (4) Socrates sought to devote reason, critical discussion, to the task of attempting to discover what is of value in life and how it is to be realized. (5) In particular, Socrates can be interpreted as putting into practice a conception of reason close to the aim-oriented rationalism advocated here, in that his basic endeavour was to subject to critical scrutiny diverse views as to what our aims in life ought to be, diverse views about such ideals as justice, goodness, courage, friendship, happiness, love. This is not, of course, the standard interpretation of Socrates' methodology. From Aristotle down to the present day, Socrates has been interpreted as holding the absurd view that in order to be virtuous it suffices to have *knowledge* of virtue, this to be acquired by arriving at correct definitions of key moral terms. The need to accommodate Socrates within the framework of the philosophy of knowledge has in this way grotesquely distorted what he actually sought to achieve: to get into personal and social life the habit of looking critically at actual and possible life aims and ideals in an attempt to improve them by rational means. Nothing could indicate more strikingly our tragic failure, even today, to have understood and taken seriously even the most elementary points Socrates sought to communicate. (6) A basic feature of Socrates' life and work, as revealed in Xenophon's Socratic writings and in the early, more historically accurate dialogues of Plato, is just the fundamentally practical, social and moral character of Socrates' concerns. Socrates did not seek primarily to make an intellectual contribution to thought (like Aristotle, and most modern academics); rather he sought primarily to make a practical, social and moral contribution to Athenian life. He sought to bring about a social and moral revolution – one which led to Socratic doubt and inquiry becoming a standard part of Athenian life. He sought to promote wisdom in life by rational means – and not mere intellectual wisdom or knowledge. (7) One kind of wisdom was recognized by Socrates to exist, namely the wisdom, the skill, of craftsmen. It is hardly too much to say that Socrates' central concern was to discover how skills employed and learned by craftsmen in creating things of value could be generalized to become skills that enable us to realize what is of most value in life. There is here a further indication of the practical character of Socrates' concerns, and an indication even that Socrates took action and the problems of action to be more fundamental than knowledge and the problems of knowledge. (8) In opposition to the Sophists, Socrates firmly rejected any mere subjectivist or relativist conception of value. What is of value is to be discovered, it is not simply what we decide. It is this rejection of subjectivism and relativism that makes Socratic ignorance possible.

Socrates might be interpreted to be arguing along the following lines: 'There is something here, implicit in our lives in Athens, that is of immense desirability and value, of profound grandeur, significance and beauty. This is to be seen in the world around us, but above all in ourselves, in our souls and in our civilization – in our crafts, our art, sculpture, poetry, drama, philosophy, in our freedom, democracy and justice. There are, however, in our souls and civilization devastating flaws – war, tyranny, injustice, violence, almost psychopathic ambition, deception, vanity and self-deception, self-annihilation of the soul. Our task is to discover how to help let that which we glimpse – of such supreme value, grandeur, significance and beauty – to grow, to come progressively into existence throughout our shared life, our *polis*. Fundamentally what we need to do is to improve appropriately our aims and methods in life – our actions, our lives, the movements of our souls. It is towards the accomplishment of this task that we need to devote our *thinking*, our rational inquiry and our education.'

It is in this way that we need to interpret Socrates' own account of his life-work:

> So long as I breathe and have the strength to do it, I will not cease philosophizing, exhorting you, indicting whichever of you I happen to meet, telling him in my customary way:

Esteemed friend, citizen of Athens, the greatest city in the world, so outstanding in both intelligence and power, aren't you ashamed to care so much to make all the money you can, and to advance your reputation and prestige – while for truth and wisdom and the improvement of your soul you have no care or worry?

CHAPTER SIX

Present Domination of the Philosophy of Knowledge in the Academic World

My claim is that it is the philosophy of knowledge – and not the philosophy of wisdom – that is the generally adopted, official view as to what ought to be the aims and methods of academic inquiry in universities throughout the world. The philosophy of know-ledge, I claim, powerfully influences almost every aspect of the academic enterprise: the aims and methods of the formal, natural and social sciences; the way different disciplines are interrelated; the way in which decisions are reached about research priorities and the funding of research; intellectual values and priorities; style and content of contributions to academic journals, monographs, textbooks, lectures and seminars; criteria adopted by editors and referees in deciding what is to be accepted and rejected for publication; academic success and failure; academic appointments and promotions; decisions concerning the awarding of academic honours, and the composition of academic elites and power groups; style and content of university degree courses; the whole way in which the academic enterprise is related to the rest of society – to industry, politics, international affairs, religion, education and so on. An academic world which upheld and sought to put into practice the philosophy of wisdom would differ profoundly, in a multitude of ways, intellectual and institutional, from what we have at present, inquiry pursued more or less in accordance with the philosophy of knowledge.

By no means all scientists and scholars accept the philosophy of knowledge in its entirety, as outlined in chapter 2. Increasingly during the decade 1972–82 aspects of the doctrine have been subjected to criticism. Furthermore, recent work of a number of people in diverse fields may well be held to be attempts to put something like the philosophy of wisdom into practice. As I shall argue in chapter 11, the intellectual/institutional revolution that I advocate in this book – from knowledge to wisdom – is already to some extent under way. Despite this, at the time of writing it is still overwhelmingly the philosophy of knowledge – and not anything like the philosophy of wisdom – which predominates over all aspects of the academic enterprise.

How, it may be asked, does the philosophy of knowledge exercise its potent influence over so many aspects of the academic enterprise? The essential point is this. The philosophy of knowledge – like any philosophy of inquiry – specifies what is to count as a contribution to inquiry, what is to be meant by intellectual progress, and in particular what is to be judged to be intellectually important. This is a matter that potently affects and concerns everyone associated with the academic enterprise, directly or indirectly. Scientists and scholars desire passionately to contribute to inquiry, to have their contributions published, accepted and valued. This passionate desire may spring from the noblest of motives: to contribute to human understanding, to help lessen human suffering or otherwise enhance the quality of human life. Or it may spring from less noble – though by no means necessarily less passionate – motives: to achieve a kind of immortality by making a lasting contribution to thought; to become famous, establish a reputation, become honoured by colleagues; to further a career or simply earn a living. In order to realize any of these ambitions, in whatever proportion of the noble and less noble, scientists and scholars are obliged to present their contributions in a form that renders them understandable and acceptable, in a form that complies with the current philosophy of inquiry (at present the philosophy of knowledge). They must do this even if they do not agree personally with the current philosophy of inquiry. Likewise editors of journals, referees and academic publishers must ensure that work accepted for publications is good, or at least acceptable, with respect to the current philosophy of inquiry. Reputations, careers, appointments, tenure, scientific and scholarly honours, entrance to scientific or academic elites, all depend on the production and publication of work which conforms to the currently adopted philosophy of inquiry. The public face of inquiry thus tends to conform to the officially accepted philosophy even if privately many individuals may have their reservations. Students, introduced to this public face of inquiry in their education, will if anything come to believe in the current philosophy of inquiry even more strongly than their predecessors (rarely if ever encountering work which fails to conform to the official philosophy, and not realizing the extent to which this conformity is the result of pressures to publish and win recognition). The lesson will be all the more powerful for being implicit. Science students do not encounter critical discussion of the aims and methods of science as a normal part of their scientific education, precisely because standard empiricism excludes such discussion from the intellectual domain of science. The philosophy of knowledge is not expounded and taught at all: it is simply presupposed by everything that is taught. Students thus come to accept the philosophy of knowledge as a result of a process that is closer to unconscious indoctrination than to education.

By means of these mechanisms, the philosophy of knowledge comes to be accepted more and more firmly and unthinkingly by scientists and scholars. It is passionately and tenaciously upheld just because it determines something of great importance to all scientists and scholars – what is to count as intellectual progress, what is to count as an important contribution to science and scholarship.

Just in case there are any doubts about the matter (which I find hard to believe), I now give some grounds for holding that the academic enterprise is conducted primarily in accordance with the philosophy of knowledge as opposed to the philosophy of wisdom.

Let us begin with the literature on the nature and purpose of universities and higher education. There is here almost universal agreement that the aim of universities both is and ought to be to create and promote knowledge. Typical quotations are the following. 'A university is a corporation or society which devotes itself to a search after knowledge for the sake of its intrinsic value' (Truscott, *Red Brick University*, 1943, p. 45). 'We think then of a university as a community of men and women engaged in a common task, namely the pursuit of knowledge' (Seeley, *The Function of the University*, 1948, p.6). 'The university is a community of scholars and students engaged in the task of seeking truth' (Jaspers, *The Idea of the University*, 1960, p. 19). 'One common theme in research activities is that all relate to the accumulation of knowledge, whether scientific or unscientific, theoretical or practical' (Corwin and Nagi, *The Social Contexts of Research*, 1972 p. 2). 'All . . . [universities] set themselves to advance learning and knowledge by teaching and research for the benefit both of their students and of the community, and in general to give students the benefits of a unversity education. Some, notably the new technological universities, also pledge themselves to pay attention to the application of knowledge for the benefit of industry and commerce' (Venables, 'The study of higher education in Britain', 1972, p. 29). '. . . it will be commonly admitted that nowadays our expectations of universities are at least twofold: they must provide training and they must foster the preservation and advancement of knowledge' (Robbins, *Higher Education Revisited*, 1980, p. 6). Some authors even defend that extreme version of the philosophy of knowledge, according to which universities should pursue knowledge exclusively for its own sake, and not in order to help solve social problems. See, for example, Nisbet, *The Degradation of the Academic Dogma: The University in America 1945–70* (1971).

In amongst this chorus of agreement one does very occasionally come across the odd discordant voice. Thus Roszak, introducing *The Dissenting Academy* (1969), remarks of the contributors to the volume that they are convinced '. . . that the proper and central business of the academy is the public examination of man's life with respect to its moral quality. It is, from first to last, the spirit of Socrates that broods over the "dissenting academy" this volume comprises' (p. 9). There is here, however, no disagreement with what is being maintained in this book. Quite to the contrary: what Roszak and his fellow dissenters argue is that most academics betray this Socratic ideal in their pursuit of academic careers and success based on the procurement and dissemination of specialized knowledge.

In order to discover in a little more detail what philosophy of inquiry at present prevails in universities, the next obvious place to look is at current philosophy/sociology of inquiry. If academic inquiry puts anything like the philosophy of wisdom into practice, the philosophy/sociology of inquiry would be, straightforwardly enough, the imaginative and critical discussion of actual and possible aims and methods of inquiry, carried on as an integral part of inquiry itself, the basic presupposition being that the fundamental intellectual aim of inquiry is to help us enhance our capacity to realize what is of value in life, help us to devote reason to developing wiser ways of life, a wiser world. There would be no dissociation between the intellectual and social aims and aspects of inquiry: intellectual aims and problems are subordinate to our fundamental aims and problems of living.

The philosophy/sociology of inquiry as it exists at present differs from all this in just the ways one would expect granted that academic inquiry proceeds, and is held to proceed, in accordance with the philosophy of knowledge. Current philosophy/sociology of inquiry is made up of (1) epistemology (2) the philosophy of science (3) the history of science (4) the sociology of knowledge (5) the sociology of science (6) the study of science policy. All these sub-disciplines accept, without question, that the basic intellectual aim of inquiry is to improve knowledge and under-standing (thus, it is to be hoped, enriching the quality of life). There is not even the faintest whisper of the idea that we might need a different kind of inquiry, a more rational kind of inquiry, devoted to enhancing wisdom. Even the most radical critics of the scientific and academic *status quo* – Feyerabend (1978), Easlea (1973), Rose and Rose (1976), Roszak (1970), Ravetz (1971) and others – all fail to argue that what we need is a more intellectually rigorous kind of inquiry devoted to enhancing wisdom. Nothing remotely approaching this possibility receives any mention or consideration whatsoever by Passmore in his book *Science and its Critics* (1978).

One does find, it is true, some criticism of some aspects of the philosophy of knowledge. Kuhn, for example, has expressed some dissatisfaction with 'the very influential contemporary distinction between "the context of discovery" and "the context of justifica-tion" ' even though he does suppose that '. . . appropriately recast . . . [it does] have something important to tell us' (1962, p. 9). Despite this, the picture of science that Kuhn paints for us is very obviously a version of the philosophy of knowledge. Indeed, in some ways it is a highly reactionary version of the philosophy of knowledge in that, for Kuhn, criticism of fundamental assump-tions has no rational role to play in normal science, a discipline only becoming authentic mature science when philosophical discussion of fundamentals is abandoned. In this way, Kuhn provides a rationale for scientists to pursue specialized puzzle-solving dissociated from all concern for philosophical and social problems of living. A basic tenet of the philosophy of knowledge is that the intellectual domain of inquiry must be decisively split off from, and shielded from being influenced by, broader social and cultural factors. For Kuhn, this is an essential feature of mature sciences. Thus he emphasizes '. . . the unparalleled insulation of mature scientific communities from the demands of the laity and of everyday life' and elsewhere asserts '. . . compared with other professional and creative pursuits, the practitioners of a mature science are effectively insulated from the cultural milieu in which they live their extraprofessional lives' (Kuhn, 1977, p. 119). He argues that '. . . the insulation of the scientific community from society permits the individual scientist to concentrate his attention upon problems that he has good reason to believe he will be able to solve' (1962, p. 164). Kuhn even comes close to endorsing the view that science should give priority to the pursuit of specialized knowledge for its own sake in passages such as the following. 'A part of normal theoretical work, though only a small part, consists simply in the use of existing theory to predict factual information of intrinsic value. The manufacture of astronomical ephemerides, the computation of lens characteristics, and the production of radio propagation curves are examples of problems of this sort. Scientists, however, generally regard them as hack work to be relegated to engineers or technicians. At no time do very many of them appear in significant scientific journals'. (1962, p. 30).

In recent years, some philosophers of science have explored the possibility that there may be a rational, if fallible, method of discovery in science (Nickles, 1980). This ought, but does not seem, to involve the rejection of standard empiricism since, as I shall argue in chapter 9, such a rational method of discovery is only possible if aim-oriented empiricism is accepted and standard empiricism is rejected.

Some other philosophers of science have called into question the value-neutrality of some aspects of science (Rudner, 1953, pp. 186; Rescher, 1965, pp. 261–76; Gaa et al, 1977, pp. 511–618); some historians of science have argued that political and ideological issues run throughout science (Graham, 1981); some others have argued for the need to develop a kind of science and scholarship devoted to socialist objectives, to helping us develop a freer, more just and more beautiful world, or to examining critically the moral life of man (Rose and Rose, 1976; Easlea, 1973). Overall, however, current philosophy/sociology of inquiry presupposes the philosophy of knowledge. Many philosophers, historians and sociologists of science may wish to reject some minor points of detail. Very few might wish to reject major aspects of the view. No one seems to advocate putting anything like the philosophy of wisdom into practice.

Failure to adopt anything like the philosophy of wisdom and aim-oriented rationalism is also strikingly apparent in the way different branches of the philosophy/sociology of science are related to science itself, and to each other.

Aim-oriented rationality, and the philosophy of wisdom, require that the philosophy/sociology of science – that is, sustained imaginative and critical discussion of actual and possible aims and methods of science – be an integral, influential part of science itself. Standard empiricism and the philosophy of knowledge require, to the contrary, that the philosophy/sociology of science, in this sense, be excluded from the intellectual domain of science, just because discussion of this kind cannot amount to contributions to knowledge, let alone contributions to empirically testable knowledge. At present, academic philosophy and sociology of science are indeed excluded from the intellectual domain of natural science, precisely in accordance with what is required by the philosophy of knowledge, and grotesquely at odds with what is required by the philosophy of wisdom. By and large, natural science just ignores academic philosophy and sociology of science.

Failure to put anything like the philosophy of wisdom into practice is also strikingly indicated by the split that exists within present-day academic philosophy/sociology of science, between the philosophy of science on the one hand, and the sociology of science on the other hand. This split mirrors the split demanded by the philosophy of knowledge between intellectual and social aspects of science. Academic philosophy of science concerns itself almost exclusively with the intellectual aspect of science. Science is presumed to have, as its basic intellectual aim, to improve knowledge of factual truth *per se*; the philosophy of science restricts itself almost entirely to considering problems that this presupposition gives rise to, such as the problem of how knowledge is possible, and the problem of what methods ought to be adopted granted we seek to acquire such knowledge. The sociology of science – a branch of sociology, and thus of social science – seeks to develop factual, scientific, sociological knowledge about science conceived of as a sociological phenomenon, an aspect of society. Both these subordinate disciplines presuppose the philosophy of knowledge. Despite this, it is almost as if they inhabit different worlds of thought between which there is scarcely any communication. Insofar as any communication does take place, it is more or less confined to ineffective, intellectual sniping. Sociologists of science can dismiss the philosophy of science for not being a part of science itself, in that it is concerned with normative questions about how science ought to proceed, and not with factual questions about how science does proceed. Philosophers of science, on the other hand, can point out dismissively that sociologists of science must presuppose some sort of philosophy of science in order to identify science itself, and in order to pursue their own discipline scientifically. This extraordinary mutual dismissiveness and lack of communication is a direct consequence of the fact that the two disciplines concern themselves with aspects of science – the intellectual and the social – between which there *ought* to be a decisive split, according to the philosophy of knowledge. The inability of philosophers and sociologists of science to speak to each other is a further striking illustration of how subtly and profoundly influential the philosophy of knowledge is on present-day academic thought.

From the standpoint of the philosophy of wisdom, this split between the philosophy and sociology of inquiry – entirely understandable granted the philosophy of knowledge – is both absurd and disastrous. According to the philosophy of wisdom, the basic task of sociology, quite generally, is to help us improve institutional aims and methods by promoting imaginative and critical discussion of actual and possible aims and methods as an integral part of the life of institutions. The basic task of the sociology of science in particular, then, is to help improve the aims and methods of science by promoting imaginative and critical discussion of actual and possible aims and methods as an integral part of science itself. This *is* the philosophy of science. According to the philosophy of wisdom, the philosophy and the sociology of science are one and the same thing: there is no distinction between them. If science is to serve humanity rationally and well, it is vital that it has associated with it imaginative and critical discussion of human or social aims for science – intellectual aims and problems being pursued and understood as subordinate to more fundamental aims and problems of people in life. This means in turn that the philosophy/sociology of science (a) is both a part of public discussion and scientific discussion (b) combines discussion of social and intellectual issues. All this is sabotaged by the current splitting up of science, philosophy and sociology of science. The sociology of science concerns itself with science as part of the human world, but cannot look critically at science from this perspective because of its concern to be factual and scientific. Thus Barnes remarks: 'It should be emphasized that the discussion is centred upon the sociologist's concern to understand and explain beliefs about nature and their variation. It does not seek to advocate or to criticize the beliefs discussed, nor is it concerned with their justification', (1974, pp. vii–ix). The philosophy of science on the other hand does leave open the possibility of criticism, since it is concerned with questions about what ought to be the aims and methods of science. Despite this, contemporary philosophy of science is quite unable to look critically at what modern science contributes to human life, just because it restricts itself to concern with the intellectual aspect of science, and seeks to portray science as a rational enterprise within the framework of the philosophy of knowledge. The net result is a general failure to discuss imaginatively and critically urgent problems concerning the capacity of modern science to help us realize what is of value in life. Such discussion cannot be an orthodox part of academic inquiry as it is at present constituted, because it does not amount to contributions to knowledge.

Do scientists themselves accept and advocate the philosophy of knowledge? The overwhelming majority of scientists do, I suggest, unquestioningly accept the philosophy of knowledge, in one or other of its versions. Despite this, full and careful formulations of the philosophy of knowledge by scientists are hard to come by, partly, I suggest, because most scientists assume the matter to be too obvious to need formulation, partly because the philosophy of knowledge puts the task of providing such a formulation outside science.

There is, however, one widely recognized spokesman for the scientific community in this respect. In contrast to the normal attitude of indifference merging into contempt that most scientists have towards the philosophy of science, the work of one philosopher of science, Karl Popper, is taken very seriously indeed by very many scientists (for example by such eminent and diverse scientists as Medawar, Bondi and Eccles). The extent of Popper's acceptance by the scientific establishment is strikingly indicated by the fact that Popper is a fellow of the Royal Society – a rare honour for a philosopher of science indeed. However, as I have already remarked in chapter 2, Popper's contributions to the philosophy of science amount to a powerful defence of one version of the philosophy of knowledge. It is very natural to interpret Popper's solution to his basic problem – the problem of demarcating science from non-science – as affirming a central tenet of the philosophy of knowledge, namely the need to restrict the intellectual domain of science to testable, factual propositions (and arguments concerning the acceptability of such propositions). Thus, the almost unprecedented way in which Popper's philosophy of science has been accepted and endorsed, almost as representing the official view, by the scientific establishment, can be taken to be a striking indication of the extent to which the philosophy of knowledge is upheld by the scientific community.

Furthermore, scientists do on occasions themselves affirm – what is generally, I claim, held to be obvious – that everything apart from testable, factual propositions is excluded from the intellectual domain of science. Thus Einstein wrote:

> . . . all scientific statements and laws have one characteristic in common: they are 'true' or 'false' (adequate or inadequate) . . . the scientific way of thinking has a further characteristic. The concepts which it uses to build up its coherent systems do not express emotions. For the scientist, there is only 'being', but no wishing, no valuing, no good, no evil – in short, no goal. As long as we remain within the realm of science proper, we can never encounter a sentence of the type: 'Thou shalt not lie'. There is something like a Puritan's restraint in the scientist who seeks truth: he keeps away from everything voluntaristic or emotional. Incidentally, this trait is the result of a slow development, peculiar to modern western thought. (1953, p. 779)

A somewhat similar affirmation of the value-neutrality of pure science has been made more recently by Professor Sir Ernst Chain, FRS, in the following terms:

> *science, as long as it limits itself to the descriptive study of the laws of Nature, has no moral or ethical quality, and this applies to the physical as well as the biological sciences.* No quality of good or evil is attached to results of research aimed at determining natural constants, such as that of gravity or the velocity of light, or measuring the movements of stars, describing the kinetic properties of an enzyme, or describing the behaviour of animals (whatever our emotional attitude towards it may be) or studying the metabolic activities of a microbe, whether harmful or beneficial to mankind, or studying physiological function or pharmacological and toxic action.
>
> No quality of good or evil can be ascribed to studies aimed at the elucidation of the chemical structure of substances of whatever nature, be it the harmless sodium chloride, the curative quinine or penicillin, or highly lethal poisons such as the botulinus toxin (a protein produced by the anaerobic *Clostridium botulinum* causing a deadly form of food poisoning, which will kill susceptible animals in amounts of fractions of a microgramme), be it the nucleic acids, the substances concerned with genetic transmission, or any other natural product, however important for life and reproduction its physiological action may be, and however potent its toxic effect on Man, animals and plants. (1970, p.166)

One version of the philosophy of knowledge insists that values are essential even to pure science – namely values that have to do with the cooperative search for truth. This position is expounded by Bronowski in his book *Science and Human Values* (1956), and later and independently by Monod in the last chapter of his *Chance and Necessity* (1974). For anyone hoping for an exposition of something approaching the philosophy of wisdom, or at least a criticism of orthodox conceptions of science, Bronowski's book seems at first sight promising. Bronowski begins by describing his arrival at Nagasaki in 1945 soon after the explosion there of the atomic bomb. It was this experience, he tells us, which prompted him to write the book. His basic problem might be put like this. What can be wrong with western science and western civilization that they can have led to the horrors of Hiroshima and Nagasaki? Any hope that Bronowski might be provoked to condemn the search for truth dissociated from the search for what is of value in life is, however, soon dashed. Bronowski merely reaffirms the traditional view that the basic aim of science is truth, the scientific search for truth making a vital contribution to civilization, the implicit assumption being that since science does have this value-neutral intellectual goal, it cannot be blamed in any way for what happened at Hiroshima and Nagasaki.

One of the most recent – and one of the clearest, fullest and most thoughtful – expositions of the philosophy of knowledge that I have come across was given by Professor Sir Andrew Huxley in his 1977 presidential address to the British Society for the Advancement of Science. Huxley sets out to defend the central tenets of standard empiricism and the philosophy of knowledge. Political, moral, religious and ideological judgements have no role to play in science; if allowed to infiltrate into the intellectual domain of science they can only serve to subvert scientific progress. The value of science resides in its capacity to acquire reliable knowledge of fact based on evidence alone. General, simplifying and unifying principles – such as the principle of conservation of energy in physics, or the principle that organisms are well adapted in biology – may provide good clues for new knowledge, but they are no substitute for evidence. In the end evidence alone, and not general principles, human hopes and fears, political, moral, religious or ideological views, must decide what is to be accepted and rejected in science. Opposition to Darwin's theory of evolution in the 1860s on religious grounds, opposition to Mendelian genetics and the adoption of Lysenko's Lamarckian views in the Soviet Union in the 1950s on ideological grounds, and more recent opposition to scientific research into questions concerning inheritance, intelligence and race in the 1970s on moral and political grounds, all constitute illegitimate and potentially damaging intrusions of human hopes and fears into the intellectual domain of science, where fact and evidence alone ought to prevail. Huxley concludes:

> My message then is that neither clues nor motives are permissible substitutes for evidence firmly based on experiment and observation. There are temptations, on aesthetic grounds, to give too much weight to broad unifying principles which deserve to be used only as clues for suggesting further inquiry, and there is another set of temptations, on moral grounds, to pay too much attention to what we hope the social consequences of our discoveries will be – a large part of the motive of most scientists in carrying out their work. Although I have drawn attention to several cases where motive, political or ideological, is impeding or distorting the advance of science, I am not despondent about the future. I believe that, at least in the western countries, there are enough people around, both scientists and laymen, who appreciate that *in the long run the value of science depends entirely on its conclusions being independent of wishes and fears about their practical application*, and who will rally to the defence of science if the pressures that I have spoken of become severe. (1977; my italics)

In the end what matters, of course, is the philosophy of inquiry that is actually put into scientific practice rather than the philosophy of inquiry that is consciously believed by scientists (by no means necessarily one and the same thing). Does science in practice embody and institutionalize the philosophy of knowledge rather than the philosophy of wisdom?

In order to answer this question, it is essential to identify the intellectual domain of science in a practical and institutional sense. The obvious way to do this is in terms of science's own identification of its intellectual domain, by means of 'science abstracts'. Not only do the various 'science abstracts' provide a record of what various scientific disciplines decide constitute contributions to science: in addition, this record plays an important role in actually shaping future science – the function of 'science abstracts' being to put scientists in touch with relevant published material.

I have therefore examined the contents of various 'science abstracts' for the year 1980.

In that year *Physics Abstracts* records 109,577 contributions to physics in published papers and books. Of these, just thirty-eight are devoted to 'philosophy of science'; and only fourteen are devoted to problems of 'science and society'. Furthermore, all the recorded contributions to the philosophy of science either presuppose or defend some version of standard empiricism and the philosophy of knowledge. Not one contribution listed in the 'Philosophy of Science' or 'Science and Society' subsections reveals the faintest glimmering of an awareness that discussion of social problems and possible and actual social goals for physics might have some relevance for the philosophy of physics. There is nowhere the faintest hint of the idea that if physics is to be both intellectually rigorous and of real human value then it is absolutely essential for the intellectual domain of physics to include some imaginative and critical discussion of actual and possible human or social goals for physics and how these influence choice of research problems in physics. Neither the technological aspect of physics, nor the cultural aspect (theoretical physics pursued as a vital part of our endeavour to improve our understanding of the universe and our place in it) show any signs of being consciously pursued in accordance with aim-oriented empiricism or aim-oriented rationalism (as characterized in chapter 5).

It must be admitted that one or two of the fourteen (out of 109,577) contributions listed in the 'Science and Society' subsection can be interpreted as developing highly specific applications of the general argument developed in this book. Thus E. Woollett summarizes his paper, entitled 'Physics and Modern Warfare: the Awkward Silence' as follows: 'General education students enrolled in courses in physics or physical science are ill-served by an almost total lack of discussion of the intimate links between progress in science and "progress" in weapons systems. The author discusses in detail the great dependence of the present arms race on a healthy physics enterprise and the pervasive connections between pure and applied science and military needs.'

General failure of physics courses to discuss the intimate links between scientific and weapons research is, I claim, one highly specific – and extremely important – example of something far more general: the failure of science to give intellectual priority to the discussion of problems of living, so that problems of knowledge are tackled as rationally subordinate to problems of living. Institutionalization of the philosophy of wisdom would automatically demand that all science courses give intellectual priority to a consideration of human problems. Institutionalization of the philosophy of knowledge automatically excludes discussion of human problems of living from the intellectual domain of science.

The contents of Woollett's paper, and of one or two other papers with related themes, powerfully confirm the basic claim of this chapter. At the same time, the mere existence of such papers, however few in number, recorded in *Physics Abstracts*, can perhaps be interpreted to be the first minute, fragmentary signs or hints of a possible future dramatic change in the overall philosophy of physics, from knowledge to wisdom. It deserves to be noted that the subsection 'Science and Society' in *Physics Abstracts* was introduced only in 1977; before that the 'General' section only had subsections: 'Communication'; 'Education'; 'History'; 'Philosophy'. In *Physics Abstracts* in 1965, there are just four papers on the philosophy of physics out of a total of 34,000 contributions to physics: no paper has a theme remotely touching on any 'Science and Society' topic.

One qualification must be made to the above. In the year 1980, *Physics Abstracts* has sixty-two main sections. So far just one of these has been discussed, namely the one entitled 'Communication, Education, History and Philosophy' (section 01.00). There are however two sections that are concerned with technology related to human problems, namely: 'Energy Research and Environmental Science' (section 86.00), and 'Biophysics, Medical Physics, and Biomedical Engineering' (section 87.00). Of course, the mere existence of technological research that is devoted to helping to solve human problems does not in itself favour either of the philosophies of inquiry. The 'Energy Research and Environmental Science' section does, however, include some contributions which explictly discuss environmental problems, and problems of energy policy. These contributions comply more with the intellectual standards of the philosophy of wisdom than with those of the philosophy of knowledge. It is however significant that this section appears to have only been introduced into *Physics Abstracts* in the year 1979; I have been unable to find any trace of the section in earlier years.

There are, in short, a very few recent contributions to the intellectual domain of physics (as defined by *Physics Abstracts*) which can be interpreted as specific, limited attempts to discuss problems of living related to physics. The great bulk of contributions to physics conforms, however, entirely to the edicts of the philosophy of knowledge. There is no hint of the philosophy of wisdom or of aim-oriented empiricism in the way the intellectual domain of physics is organized, into sections and subsections, as depicted by *Physics Abstracts*.

The few, marginal, scattered hints of some aspects of the philosophy of wisdom that are to be found in a few contributions listed in *Physics Abstracts*, disappear altogether, however, when one turns to *Chemical Abstracts*. Here there is no philosophy of science subsection and no science and society subsection: instead there is one section (out of a total of thirty-four) entitled 'History, Education, and Documentation'.

Abstracts for other disciplines in the physical sciences and technologies by and large confirm this general picture. Somewhat exceptional, perhaps, are *Electrical and Electronic Abstracts*, with such subsections as 'Administration and Management' and 'Planning', and *Computer and Control Abstracts*, with such subsections as 'Philosophical Aspects', 'Economic, Social and Political Aspects', and sections such as 'Systems Theory and Cybernetics' and 'Administrative Data Processing'.

As one moves from the physical sciences to the biological, medical and human sciences, so, as one might expect, more and more concessions are made to the need to discuss human problems of living. There are few signs of deviation from the philosophy of knowledge in *The Zoological Record*; in *Biological Abstracts*, however, there are sections which include contributions that discuss social problems, such as 'Public Health' and 'Psychiatry'. *Geo Abstracts* devotes entire sections to contributions to geography that are, in one way or another, concerned with human problems: for example, 'Economic Geography C', 'Social and Historical Geography D', and most strikingly 'Regional and Community Planning F' – many contributions to geography listed in this section being concerned with just the kind of problems at the centre of this book. *Psychological Abstracts* too, not surprisingly perhaps, includes, amongst its sixteen sections, some that are concerned with problems of living – such as 'Social Processes and Social Issues', 'Educational Psychology' and 'Applied Psychology' (although one cannot help but note that the very term 'applied' as used here, presupposes the philosophy of knowledge). In the fields of psychiatry, government, political science, economics, sociology and international affairs, it is of course standard for social problems, of one kind or another, to receive attention. There are academic journals in these fields more or less devoted to the discussion of social problems of one kind or another: for example, *Applied Economics, Development, Economic Impact, Human Relations, International Affairs, International Relations, Journal of Black Studies, Journal of Conflict Resolution, Journal of Development Studies, Journal of Social Policy, New Society, Political Science Quarterly, Radical Philosophy, Science and Society, Social Policy and Administration, Social Problems.*

One might be inclined to conclude from this list of academic journal titles alone, that the various branches of social inquiry at least – economics, sociology, psychology, political science, the study of international affairs and anthropology – do put the philosophy of wisdom into practice (if only the version outlined in chapter 4). In fact, quite to the contrary, these academic disciplines all presuppose the philosophy of knowledge (in one or other of its versions). This holds even when, as in the case of economics or psychiatry, the basic rationale for the discipline is clearly recognized by everyone to be to help solve a group of important human problems of living.

Thus modern economics from the outset, with the work of James Steuart, Adam Smith, Malthus, Ricardo and J.S. Mill, conceived of itself, in the spirit of the Enlightenment, as a *science*, a discipline devoted to the acquisition of *knowledge* about economic phenomena which, when applied, would help solve practical economic problems (or show that some cannot be solved). Even Marx upheld this conception of economics. In the preface to *Das Kapital*, Marx makes it quite clear that he is making a contribution to knowledge, to the science of political economy. He tells us 'It is the ultimate aim of this work to lay bare the . . . law of motion of modern society' (1921, I, p. 14) and he quotes with approval the following remarks of a reviewer of an earlier edition of the book: 'Whilst Marx sets himself the task of following and explaining from this point of view the economic system established by the sway of capital, he is only formulating, in a strictly scientific manner, the aim that every accurate investigation into economic life must have. The scientific value of such an inquiry lies in the disclosing of the special laws that regulate the origin, existence, development, and death of a given social organism and its replacement by another and higher one' (p. 24). This tradition continues through the work of figures like Jevons, Menger, Marshall and Keynes down to the present day. There are, of course, disagreements and developments within economics: but these are disagreements about, and developments in, what is taken to be theoretical economic *knowledge*. Any standard history of economic thought – such as Blaug (1968) – is a history of economic *science*, a history of the endeavour to improve knowledge of the economic aspect of life. J.N. Keynes, father of *the* Keynes, in his introductory remarks to his *Scope and Method of Political Economy* (1890), makes it quite clear than he subscribes to this general view. Thus he remarks *'Political economy* or *economics* is a body of doctrine relating to economic phenomena . . . the purpose of the following pages is to discuss the character and scope of this doctrine, and the logical method appropriate to its development. In seeking to define the *scope* of any department of study, the object in view is primarily to determine the distinguishing features of the phenomena with which it deals, and the kind of knowledge that it seeks concerning these phenomena' (p. 2). Friedman again asserts 'Positive economics is in principle independent of any particular ethical position or normative judgements. As Keynes says, it deals with "what is", not with "what ought to be". Its task is to provide a system of generalizations that can be used to make correct predictions about the consequences of any change of circumstances' (1968, p. 509). Robbins, in his significantly titled *An Essay on the Nature and Significance of Economic Science*, ponders, in the first chapter, the problem of what it is that economic science seeks to acquire knowledge about, the assumption that economics does seek knowledge being so entirely taken for granted that it is not even explicitly stated. In the second chapter Robbins remarks '. . . Economics is entirely neutral between ends; that is, in so far as the achievement of any end is dependent on scarce means, it is germane to the preoccupations of the economist. Economics is not concerned with ends as such' (1952, p. 24). Joan Robinson, more recently, remarks 'Economics . . . consists . . . of imperfectly tested hypotheses – about how an economy works, why one economy differs from another, what consequences are to be expected from any particular events or particular policies' (1960, pp. xv–xvi). Worswick more recently still, in an essay entitled 'Is progress in economic science possible?' remarks '. . . we conceive progress in economic science as consisting of a dialogue, or interaction, between fact and theory, the latter being strengthened or modified according as new data come to light and according as they agree or disagree with hypotheses deduced from theory' (1972). Worswick makes clear that 'theory' here is to be understood as factual, explanatory theory 'as in any science'.

Finally Hollis and Nell in their *Rational Economic Man* remark, 'we share the view advanced here that economic theory is a branch of a more general tree of knowledge' (1975, p. 251).

There are signs of a growing sense of malaise amongst economists about the nature and status of their discipline – due in part to an awareness of recalcitrant economic problems in the real world, and also to a more general feeling of unease amongst social scientists about the nature and success of social inquiry. This sense of malaise finds expression in books such as Ward, *What's Wrong with Economics* (1972); Hutchison, *Knowledge and Ignorance in Economics* (1977); Bell and Kristal (eds), *The Crisis in Economic Theory* (1972). I have hunted in vain, however, in the economic literature, for signs of the philosophy of wisdom being advocated and consciously put into practice in economic inquiry. The task of helping us to put aim-oriented rationality into practice in our economic activities in a cooperative and just fashion, so that we all benefit, is not the central intellectual concern of economic inquiry. Methodology and philosophy are discussed by economists far more seriously than by natural scientists: see for example, works by von Mises (1960), Hutchison (1978) and Blaug (1980). But what is invariably discussed is the methodology and philosophy of *economic inquiry*, conceived to have the basic intellectual aim of acquiring knowledge of economic phenomena, of use for economic activity itself. What economists do not do is to endeavour to apply the methodology and philosophy of aim-oriented rationalism to economic activity itself, to economic pursuits – economic inquiry concerning itself with the problems that such an endeavour gives rise to. This disastrous intellectual and professional failure of economists during the last two centuries is not unrelated to the very serious economic problems that confront humanity today – wealth and poverty distributed amongst people and nations in an appallingly unjust way, rapid depletion of capital in the form of finite natural resources, world recession, unemployment, the failure to develop the free market system so that it functions in such a way that we all enhance our capacity to realize what is genuinely of value to us in life.

Analogous states of affairs prevail in the other branches of social inquiry – in sociology, psychology, history, theoretical psychiatry and psychotherapy, political science, anthropology, the study of international affairs, development studies, conflict studies and so on. In all these fields, the basic intellectual aim is to improve factual knowledge – and understanding – of different aspects of social phenomena, of the social world. This central intellectual aim goes back to the birth of the discipline, even when the discipline can be traced as far back as the Enlightenment of the eighteenth century. Social inquiry as science, as inquiry with the basic aim of improving knowledge of the human world, was indeed in a sense the creation of the Enlightenment. Almost all those in the eighteenth, nineteenth and twentieth centuries recognized by academic social inquiry itself to have made the most important contributions to the field have advocated, and have sought to put into practice in their work, some version of the philosophy of knowledge. This is true, for example, of such diverse major figures in sociology as Montesquieu and Ferguson in the eighteenth century, Comte, Mill and Marx in the nineteenth century, Pareto, Durkheim and Weber in the late nineteenth and early twentieth centuries. It is true of more recent writers in the field, such as Talcott Parsons, Alfred Schutz, R. Nisbet, Raymond Aron, Erving Goffman, Donald MacRae, John Rex, Alan Ryan and Anthony Giddens. In psychology it is true of such diverse figures as Wundt, James, Galton, McDougall, Fechner, Köhler, Koffka, Wertheimer, Pavlov, Watson, Piaget, Burt, Skinner, Eysenck, Hudson.

In all these diverse fields of social inquiry, both classic texts and introductory textbooks, assume or assert that the discipline has, as its basic intellectual aim, to improve knowledge and understanding of the relevant aspect of the human world. Thus in sociology, a major figure such as Weber declares that 'Sociology . . . is a science which attempts the interpretative understanding of social action in order thereby to arrive at a causal explanation of its course and effects' (1947, p. 80). The following quotations are typical of the introductory remarks to be found in elementary sociology textbooks. 'Sociology is one of the social sciences. Its purpose is the scientific study of human society through the investigation of the social behaviour of man' (Giner, 1972, p. 9). 'Sociology is the scientific study of human interaction. It is also the body of knowledge about human interaction resulting from such study' (Cairns and Dressler, 1973, p. 3). 'As a scientific field, sociology is both academic and applied. Like all scholars, sociologists try to be exact. As a result, they have developed ways of studying social life that may seem remote from urgent human concerns. This does not mean that sociologists do not care. Most want to help solve social problems, but if research is to be socially useful, it must be sound and objective. Thus, the sociologist is pulled in two directions: towards careful, well-designed studies and toward efforts to solve pressing social problems. The tension between these two priorities is expressed in a debate that has been going on for more than a generation' (Broom *et al.*, 1981, p. 7).

Analogous quotations can readily be accumulated from the classic works, and from the introductory textbooks, of the other fields of social inquiry – psychology, anthropology, history, political science, and so on.

As in economics, and as in sociology (as the last quotation indicates), so too in the other fields of social inquiry there is a general concern to help solve personal and social problems of living experienced by people in life, social inquiry seeking, in this way, to contribute to the promotion of human welfare. The fundamental idea that prevails, however, throughout the diverse branches of social inquiry, is that the proper way for each discipline to contribute towards the resolution of personal and social problems of living is, in the first instance, to acquire relevant factual, theoretical and explanatory knowledge, and then apply this knowledge to helping to solve social problems. A distinction is maintained between *social* problems and *intellectual* problems of knowledge and understanding of each discipline – the primary intellectual task of each discipline being to solve these latter problems of knowledge and understanding. As one sociological textbook remarks 'Social problems are not the same thing as sociological problems' (Worsley *et al.*, 1970, p. 51). All this exemplifies the philosophy of knowledge, and echoes the analogous situation to be found in the physical and biological sciences (such as the distinction between technological and applied scientific research on the one hand, pure scientific research on the other hand).

Social 'scientists' do of course disagree amongst themselves about the relative merits of 'pure' and 'applied' social inquiry; they disagree as to where priorities ought to lie, in terms of money and research effort. Furthermore, they disagree in their views as to what the most urgent and important human problems are in the world today, and what needs to be done in order to help resolve them. Some social 'scientists' are politically conservative, others uphold liberalism, others democratic socialism, and others again some form of revolutionary socialism. All this echoes analogous states of affairs to be found in the physical, biological and technological sciences, and corresponds to what one would expect to find, granted that those fields are dominated by the philosophy of knowledge.

Another major long-standing disagreement that runs through most branches of social inquiry – and one which has more to do with the nature of social inquiry itself – is the disagreement between the pro-naturalists and the anti-naturalists. Pro-naturalists such as Popper, Friedman, Blaug, Harris, Hutchison, Skinner, Eysenck, Wilson, Keat and Urry, can trace the naturalist tradition in diverse branches of social inquiry back to the Enlightenment. Anti-naturalists of one kind or another – Winch, Laing, Berlin, Goffman, Giddens, Shotter, Habermas, Foucault, Gadamer, Betti – can trace their tradition back equally far, to what Berlin has called the Counter-Enlightenment. Via herme-neutics, phenomenology and existentialism – via the writings of such figures as Schutz, Collingwood, Merleau-Ponty, Sartre, Heidegger, Husserl, Nietzsche, Kierkegaard, Dilthey, Schleier-macher – this tradition can be traced back to early nineteenth- and late eighteenth-century Romantic thought and literature, to Herder and to Vico. However, insofar as this long-standing debate is about whether or not the kind of knowledge and understanding to be sought in social inquiry, and the kind of methods to be employed, are similar to the knowledge, understanding and methods of the natural sciences, this debate is internal to the philosophy of knowledge. It is a debate about what version of the philosophy of knowledge to adopt for social inquiry. The existence of the debate is itself a striking confirmation of the thesis that the philosophy of knowledge – in one or other version – prevails in social inquiry. For, of course, from the standpoint of the philosophy of wisdom, the debate disappears.

According to the philosophy of wisdom, the aim of all of inquiry – including natural science – is to help us to live more aim-oriented rationalistic lives and develop more aim-oriented rationalistic institutions. Philosophy/sociology of science is a special case of this. Social inquiry is social methodology, very similar to scientific methodology, but of course quite unlike scientific inquiry itself. There is unity of aim and method throughout all of inquiry even though social inquiry is quite different from natural science. The philosophy of wisdom reveals with striking clarity that pro-naturalists and anti-naturalists are both partly right and partly wrong, and that the debate itself is entirely misconceived. The persistence of the debate amongst scholars pursuing social inquiry indeed helps to maintain the philosophy of knowledge by distracting attention away from the vital need to pursue social inquiry as reason devoted to the growth of wisdom in life, and by creating artificial divisions, conflicting intellectual interests and concerns, so that social inquirers fail to engage in the massive *cooperative* venture of putting the philosophy of wisdom into practice throughout all of social inquiry – indeed throughout all of inquiry and of life.

There is, it is true, a great deal of methodological discussion in social inquiry, not only in economics, but also in sociology, psychology, anthropology and elsewhere. But this methodological discussion is almost entirely devoted to the problems concerning the methodology of the diverse branches of social inquiry conceived to have the basic intellectual aim of improving knowledge and understanding of aspects of the human world. It is not at all concerned with problems arising from the endeavour to put aim-oriented rationalistic methodologies into practice in our diverse personal, institutional and social enterprises in life. It thus presupposes versions of the philosophy of knowledge throughout, and fails disastrously to be social inquiry pursued in accordance with the philosophy of wisdom.

The philosophy of knowledge has such a vice-like grip on the minds of social inquirers, that on occasions it leads them to project it onto the human world, and even onto the biological world. Thus Kelly, sensibly and correctly enough, recognizes that people pursue inquiry in their lives: but because the philosophy of wisdom is not available to him, he cannot make the elementary points made here – namely that personal thinking is fundamentally concerned with problems of action, with realizing various desired goals in life, being pursued, ideally, in an (aim-oriented) rational way – all academic inquiry having, as its basic rationale, to promote personal inquiry in life so that we may realize what is of value to us. Kelly, gripped by the philosophy of knowledge, is obliged to interpret the personal inquirer as a sort of scientist, seeking knowledge. Personal construct theory then itself seeks to develop academic psychological knowledge about the knowledge acquiring, or construct building, endeavours of individuals (Kelly, 1955; Bannister and Fransella, 1971).

A second example of the phenomenon is to be found in the interpretation of evolution recently put forward by Plotkin and Odling-Smee (1981). These authors, quite properly, deplore the tendency of some workers on the theory of evolution, such as Dawkins (1978), to interpret evolution primarily in mechanistic or physical terms – so that the theory of evolution is used to explain purposiveness away, rather than to help us understand how and why purposiveness has gradually evolved in the world. All this is excellent. But then Plotkin and Odling-Smee go on to make the disastrous assumption that the goal of life can be taken to be knowledge-gain, it thus being legitimate to interpret evolution as the evolution of knowledge, learning being interpreted, in philosophy of knowledge terms, as acquiring knowledge, and not, in philosophy of wisdom terms, as learning how to live. It is almost as if the authors conceive of all of life as striving to do what academic scientists and scholars do – acquire knowledge. In fact, of course, though knowledge-acquisition is implicated in animal and human living and learning, both animals and humans pursue many goals, and learn how to pursue many goals besides that of acquiring knowledge.

As in economics, so in other branches of social inquiry, there is a constant sense of malaise, a sense that there is something fundamentally wrong in the way each discipline is pursued and conceived: this is, for example, indicated in the very titles of works such as: Gouldner, *The Coming Crisis in Western Sociology* (1970); Joynson, 'The breakdown of modern psychology' (1970); Brown (ed.), *Radical Psychology* (1973); Dyal *et al.*, *Readings in Psychology: the Search for Alternatives* (1975); Heather, *Radical Perspectives in Psychology* (1976), all referred to in Westland, *Current Crisis of Psychology* (1978). So far, however, this sense of crisis has not led to a general recognition that all this is but part of the general crisis that confronts humanity today due to its long-standing failure to put the philosophy of wisdom into practice in personal and social life – its long-standing failure, indeed, even to conceive of the urgent need to attempt to do this.

The central, and tragic, intellectual mistake (according to the philosophy of wisdom) that has bedevilled social inquiry ever since the Englightenment is illustrated in miniature in an especially graphic and simple way in a book by Barbara Wootton entitled *Testament for Social Science* (1950). Its subtitle – 'An Essay in the Application of Scientific Method to Human Problems' – might lead one to believe that the book expounds and defends the philosophy of wisdom. But if the first few sentences of the book strengthen this belief, what follows must quickly dispel the idea. The book opens as follows:

> The contrast between man's amazing ability to manipulate his material environment and his pitiful incompetence in managing his own affairs is now as commonplace as it is tragic. The world of atomic energy and nylon is for millions still the world of poverty, hunger, misgovernment, crime, domestic unhappiness or personal frustration. And mastery over earth and air and sea and atom has brought us only to daily fear of sudden death of our own making. No one has any doubt how that mastery has been won. It is by vigorous

devotion to scientific method that we have made our conquests over the material environment . . . It is no less obvious that this method, which has been so brilliantly successful in the natural sciences, is not normally applied to the field of our most disastrous failures. The personal relations of human beings, individual and collective, are conducted in a quite different way: these are for the most part governed by a medley of primitive impulses, kindly or harsh, sometimes even noble, modified by rules of thumb, and set in a framework of a traditional morality which varies from place to place and from age to age. In these matters science plays little part and commands but meagre respect . . . experience (falls) into two sharply divided sections – that in which science speaks with authority, and that in which she whispers furtively, or is dumb. This contrast surely seems to point a simple moral – that one ought seriously to ask whether the tool that has worked such wonders in the one job could not be used for the other. More than a century has passed since Auguste Comte said that the rational reform of society must be brought about by the application of scientific method to social problems. If not very much has happened since to prove him right, certainly even less has happened since to prove him wrong. In the intervening century scientific method has marched from victory to victory in the field of natural phenomena, while those human problems which have not enjoyed its attention remain as intractable as ever. It is, therefore, the first purpose of this esssay to ask how far these problems also might be tackled by the methods of science . . I hope to show (that) the potential contribution of science in this field is far greater than anything we have yet seen: the differences between the material of the social and the natural sciences are differences of degree, rather than of kind. (Wootton, 1950, pp. 1–3)

Any lingering doubts one might have as to what Wootton is advocating vanish altogether with her second chapter entitled 'Scientific Method in the Social Sciences'. The chapter begins: 'The stages of the scientific progress are now generally familiar. There is first the accurate observation of data; then the formulation of an hypothesis; and finally the promotion of the hypothesis by empirical verification to the status of a law. Scientific method "is simply the attempt to acquire knowledge of general laws directly or indirectly by experience, by the use of our five senses" (A.D. Ritchie, *Scientific Method*, Kegan Paul, p. 189). Our problem is thus to determine how far a parallel attempt can be made to acquire knowledge about human relationships (1950, p. 6).

In short, Wootton is not advocating that methods that have proved successful in the cooperative endeavour of science be generalized and exploited in our other cooperative endeavours in life, with their diverse aims – government, industry, art, literature, marriage, love, and so on. She is not arguing that social inquiry be pursued as social methodology – with the task of helping us to develop (aim-oriented) rationality in life. In absolute contrast to this, she assumes, without a flicker of doubt, that the task of applying 'scientific method to human problems' involves applying scientific method to *the enterprise of improving knowledge about social phenomena* – the central task being to develop social science on analogy with natural science. From the standpoint of the philosophy of wisdom, this is perhaps the basic intellectual disaster of the philosophy of knowledge.

Finally, the philosophy of knowledge has exercised a profound influence over the entire field of modern *philosophy*, from Descartes to the present. Indeed, one might almost say that the central problems of philosophy, in this tradition, have been problems posed by 'the philosophy of knowledge'. What can we know? How can we acquire knowledge – whether common-sense knowledge or scientific knowledge? How are arguments which seem to show we cannot acquire knowledge to be refuted? Of what can we be certain? What methods need to be employed in order to improve knowledge? In terms of what criteria do we assess the progress of scientific knowledge? Can philosophy provide us with a special kind of non-scientific metaphysical – or phenomeno-logical – knowledge? What are the limits of the knowable? These epistemological, methodological and metaphysical problems – and associated problems to do with perception, causation, the relationship between the mind and the brain, knowledge of the past and of other minds – may well be held to be the central problems of philosophy since Descartes. Those thinkers generally held to have made the most substantial contributions to this tradition of philosophy are generally understood to have been centrally preoccupied with these problems: Bacon, Descartes, Locke, Spinoza, Leibniz, Berkeley, Hume, Kant, Mill, Whewell, Bolzano, Brentano, Husserl, Mach, Bradley, McTaggart, Frege, Peirce, James, Moore, Russell, Whitehead, Poincaré, Meyerson,

Cassirer, Duhem, von Mises, Campbell, Hanson, Polanyi, Lewis, Schlick, Reichenbach, Carnap. The same holds for more recent philosophers: Hempel, Nagel, Ayer, Popper, Ryle, Strawson, W. Kneale, Quine, Grünbaum, Körner, Feigl, J.J.C. Smart, Kuhn, Agassi, Lakatos, Watkins, Hesse, Harré, Shimony, Madden, Salmon, Sellers, Suppes, Hacking, Toulmin, Black, Putnam, Levi, Mackie, Quinton, Scriven, Feyerabend – and many others. There are of course exceptions. Eighteenth-century philosophers such as Voltaire, Diderot, Condorcet, Paine, Godwin and Wollstonecraft, passionately concerned to devote reason to the growth of enlightenment – to the growth of tolerance, justice, happiness, love, individual liberty, democracy – do not fit very well into the general picture of philosophy as a part of the pursuit of knowledge: their role in the history of western philosophy is appropriately downgraded. Machiavelli, Hobbes, Rousseau, Hegel, Schopenhauer, Marx, Kierkegaard, Nietzsche, Wittgenstein, Sartre, Burtt, Hayek do not perhaps, in their very different ways, entirely conform to the general pattern. On the whole, however, philosophy is centrally concerned with problems of knowledge; and even where other branches of philosophy are pursued – moral, political, aesthetic, religious or educational – nevertheless the central intellectual concern remains to make a contribution to knowledge.

All this is quite startling when one considers that 'philosophy' traditionally means 'the love of wisdom'. Modern philosphy, entirely self-consciously, stems from ancient Greece, from Socrates, his contemporaries, predecessors and successors, most notably Plato and Aristotle. For Socrates and his contemporaries, philosophy was understood to be 'the pursuit of wisdom'. However, with Plato, and increasingly with Aristotle, 'wisdom' becomes 'knowledge'. It is this aspect of ancient Greek philosophy that became prominent, in the sixteenth and seventeenth centuries, with the birth of 'natural' and 'experimental' philosophy – that is, science. This tendency is continued with the development of social science from the eighteenth century to today. Ancient Greek philosophy – pursued as the love of wisdom – is transformed, from the seventeenth century onwards, into the scientific pursuit of knowledge, and academic philosophy, increasingly, is confronted with the problem of discovering how it can make some contribution to the general pursuit of knowledge. Over the centuries, academic philosophy has seemed to become increasingly impoverished, as whole areas of 'philosophy' have departed to become respectable empirical sciences: physics, astronomy, sociology, psychology, logic, economics, political science, cosmology, linguistics.

It is of course the central thesis of this book that all of inquiry – mathematics, natural science, social science, technology, scholarship – needs to give intellectual priority to the task of promoting the growth of wisdom in the world. In order to be of maximum intellectual rigour and value, and of maximum human value, all inquiry needs to develop, and be an institutionalization of, what Socrates did, advocated, lived and died for. The terrible human disasters of the past and present are intimately linked with the great intellectual disaster involved in developing cooperative inquiry in such a way that intellectual priority is given to the growth of *knowledge*, rather than the growth of *wisdom in life* – an intellectual disaster that can be traced back to Descartes, to Francis Bacon, and to Aristotle and Plato.

All of those of us who are in one way or another involved with modern science, technology, scholarship or education must take some share of responsibility for the persistence of this intellectual and human disaster, that is such a fundamental and pervasive aspect of the modern world. Those of us who are in one way or another involved with academic philosophy, however, bear an especially heavy burden of responsibility. For the disaster is basically a *philosophical* disaster, a persistent misrepresentation of what ought to be the basic aim and methods of all rational inquiry. It is the central professional concern of academic philosophy to develop a good generally acceptable philosophy of rational inquiry – a view as to what the basic aim and methods of inquiry ought to be. What academic philosophy ought to have done, over the decades and centuries, building on what is best in the life and work of Socrates and the Sophists, and the 'philosophes' of the Enlightenment, is to have advocated that all of inquiry needs to give intellectual priority to the growth of wisdom. Furthermore, academic philosophy should have done everything in its power to help develop all of science, technology, scholarship and education in this way. And not content with this, academic philosophy should have sought to help promote imaginative and critical discussion of aims and methods in all other human endeavours as well – politics, industry, law, the media and so on – thus helping us quite generally to put cooperative aim-oriented rationalism into practice in personal, social and institutional life, so that we may all the better realize what is of value to us as we live. From this standpoint, of course, philosophy and social inquiry are one and the same thing: just as the philosophy and the sociology of inquiry are identical, so too are economic philosophy and economic 'science', political philosophy and political 'science', social philosophy in general, and sociology, psychological philosophy and psychological 'science'. All these branches of philosophical/social inquiry are fundamentally *methodological* in character, and need to be an integral, influential part of that aspect of life with which they deal (in order to promote aim-oriented rationality, and wisdom, in life). As a result of being developed in this way, academic philosophy would have had an important, fruitful contribution to make to modern science and scholarship, and to modern life.

Instead of all this, academic philosophy, by and large, has taken it for granted that the basic intellectual aim of all of inquiry – and of philosophy in particular – is to improve knowledge. And as a result, in part, we have failed all too often to develop traditions of improving personal, social and institutional aims and methods towards the realization of what is of most value to us in life. We have failed to develop organized inquiry in such a way that it gives intellectual priority to the task of promoting wisdom in life.

All of life, and all of inquiry, has suffered to a greater or lesser extent as a result of the intellectual failure of philosophy to give priority to the task of promoting wisdom in life. And incidentally, as it were, academic philosophy has suffered as well. For whereas as the pursuit of *wisdom* philosophy has much to offer, as the pursuit of *knowledge* philosophy is reduced to an absurdity or a triviality. In the first place, it cannot solve its fundamental problem – the various aspects of the problem of knowledge. For, in order to solve this problem, it is essential to construe the pursuit of knowledge as an aspect of *life*, an aspect of the pursuit of wisdom in life. It is essential to give intellectual priority to action, to life, and to the capacity to act more or less successfully in the world. Traditional philosophical problems of knowledge – such as the problem of induction, the problem of the rationality of science – have resisted resolution precisely because these problems cannot be resolved within the framework of the philosophy of knowledge. They are indications of the *irrationality* of this framework. Only within the framework of the philosophy of wisdom can they be resolved. In the second place, philosophy fails to make any significant contribution to knowledge. For there is no peculiarly philosophical kind of knowledge distinct from scientific or common-sense knowledge: the attempt to provide such distinct knowledge leads either to absurdities, as in Hegel, Bradley, McTaggart or early Wittgenstein, or to sterile trivialities, as in much ordinary language philosophy and conceptual analysis.[1] Furthermore, the attempt to pursue philosophy as a branch of knowledge sabotages the one thing of great value philosophy might offer – help with the rational growth of wisdom in life. Precisely in order to retain the status of a branch of knowledge, academic philosophy of morality, of politics, of art, of religion, of science, of education and so on, is obliged to draw a sharp distinction between itself, a meta-discipline, and the enterprise it studies. Thus, in this vein, Melden, introducing a textbook on moral philosophy or ethics, remarks: '. . . theoretical interest in the subject matter of ethics, whatever the conditions of its origin may be, must not be confused with the practical interest of moral beings. The theoretical interest is concerned with knowing; the practical interest is concerned with doing' (1960, p. 3). Again, Quinton, introducing a collection of papers on political philosophy, remarks: 'A comparatively definite place has now been marked out for philosophy within the total range of man's intellectual activities . . . Very briefly, philosophy has the task of classifying and analysing the terms, statement and argument of the first-order disciplines' (1968, p.1). This sharp division between 'philosophy of X' and 'X', whatever human endeavour X may be, made in the interests of philosophy being a contribution to knowledge, annihilates at a stroke the only thing of value philosophy can have to offer, namely to help build aim-oriented rationalism into X itself. For, in order to do this, it is vital that the philosophy of X – the enterprise of articulating and criticizing actual and possible aims and methods for X – must be an integral, influential part of X itself.

All of inquiry, as a result of being deemed to pursue knowledge, can be regarded as suffering from rationalistic neurosis, with all the attendant defects this state of affairs incurs, discussed in chapter 5. Rationalistic neurosis is however especially acute in academic philosophy. From the standpoint of the 'philosophy of wisdom' indeed, academic philosophy pursued as a branch of knowledge might almost be characterized as a depository of neurotic problems that arise as a result of failing to give intellectual priority quite generally to the growth of wisdom. During the course of this book I hope to show how major philosophical problems which have long resisted resolution, and

---

[1]For criticism of ordinary language philosophy and conceptual analysis, see Popper (1969, vol. 2, pp. 9–21; 1959, pp. 15–23; 1963, pp. 66–96); Gellner (1959); Maxwell (1976b, pp. 31–51).

which have not led to fruitful work, are either solved or are at least transformed into fruitful problems as a result of being set within the context of the philosophy of wisdom: for example, the problem of induction, the problem of free will and determinism, the problem of mind and brain, the problem of fact and value.

CHAPTER SEVEN

Assessment of the Basic Argument

During the twentieth century mankind has made extraordinary progress in scientific knowledge, and in technological and industrial development. During the same period, mankind has committed horrifying crimes against itself, in that millions upon millions of people have suffered and died as a result of war, tyranny, concentration camps, mass executions, economic exploitation and increasingly unjust distribution of the world's resources. A major reason for this glaring discrepancy between what has been achieved in knowledge and in life is that during the last two or three centuries – and especially during the twentieth century in the developed world – mankind has succeeded only in developing socially influential organized inquiry in accordance with the philosophy of knowledge, and has thus failed to develop organized inquiry in accordance with the philosophy of wisdom. As a result, specialized knowledge has flourished, but social wisdom in the world has faltered. If we are to progress towards a wiser world it is essential that science, technology, scholarship and education in schools, universities and research establishments throughout the world be transformed to accord with the edicts of the philosophy of wisdom. If organized inquiry is developed in this way, then we may reasonably hope to make gradual progress towards a more just, humane, cooperative – and even loving – world. This, in outline, is the central argument of this book.

I put this argument forward in all seriousness, in the hope that it will be taken up and used to help change the actual institutional structure of academic inquiry, from knowledge to wisdom. My intention is to help establish a new intellectual/institutional orthodoxy in which the philosophy of wisdom is taken for granted and built into reseach aims and priorities, into intellectual values and ideals, into criteria for publication and acceptance, into teaching and administration, into funding, appointments and careers, in universities and research establishments throughout the world. This is a tall order indeed. What I am proposing will be fiercely resisted – or will be, much more effectively, blandly ignored – for a variety of good and bad reasons. For I am advocating nothing less than that the basic aims and methods, the whole character, of the academic enterprise be changed.

It might be thought that nothing very disturbing or threatening can come from a field that is as abstract and theoretical as the philosophy of inquiry. But this is only true as long as the philosophy of inquiry restricts itself to attempting to provide a rationale for the *status quo* – some version of the philosophy of knowledge – as it traditionally has done. The moment the philosophy of inquiry comes up with radically new proposals as to what the basic aims and methods of inquiry ought to be, inevitably such proposals will be held by many to be threatening indeed and will meet with fierce resistance. For, as I explained in the first section of the last chapter, the philosophy of inquiry that is in practice accepted and built into the institutional structure of the academic enterprise, determines something that is of importance to everyone, but above all of immense importance, of passionate concern, to every scientist and scholar – namely what is to count as intellectual progress, what is to count as a contribution to inquiry, and what is to count as intellectually important. This is something that is of passionate concern to all scientists and scholars not only for the very noblest of reasons, but also for reasons that are somewhat less noble: scientific and scholarly reputations, membership of scientific and scholarly elites, academic careers and appointments, academic prizes and degrees, all depend on what the academic community in practice judges to constitute important, or acceptable, contributions to inquiry. Advocate a change in the basic intellectual aims and methods of inquiry, a change in the standards used to judge the intellectual importance or acceptability of contributions to inquiry, and one advocates something that threatens to annihilate established scientific and scholarly reputations, cancel the importance of lifetimes of scientific and scholarly work, alter the rules of the game whereby Nobel prizes are won, professorships are acquired, careers advanced, degrees attained. It is not to be wondered at that any proposal along these lines, however intellectually obvious and urgently needed it may be, will meet with fierce resistance.

There are other sources of resistance as well. The philosophy of knowledge is built into the habits of thought and work of countless scientists and scholars; and it is built into the institutional structure of the academic enterprise. In order to put the philosophy of wisdom into practice, it is just these habits of thought and work, and these institutional realities, which will need to be appropriately changed. But, of course, habits of thought and work, built up over a lifetime, are notoriously difficult to change, and become all the more difficult the older one becomes. Institutions, again, are notoriously difficult to change. This is especially true of academic institutions, as anyone who has attempted to bring about any institutional change in universities, however minor, will know only too well.

If I am to make any headway against this massive wall of resistance, I must formulate the basic arguments in support of the philosophy of wisdom, and against the philosophy of knowledge, in an overwhelmingly obvious, simple, clear and decisive fashion. This I now strive to do. If, in doing this, I labour the obvious, I apologize in advance.

There are two additional points to be taken into account. The first was briefly indicated in chapter 2. Standard empiricism and the philosophy of knowledge, as a result of being already built into the institutional structure of science, and of academic inquiry more generally, tend to exclude from the intellectual domain of discussion precisely the kind of arguments developed in this book, even though these arguments are entirely valid. Thus, according to standard empiricism, essentially only empirically testable ideas can enter the intellectual domain of science. In this book I put forward the idea that this standard empiricist conception of scientific rigour is seriously defective, as a result of misrepresenting the basic aim of science: this idea is, however, not itself testable, and thus will be excluded by standard empiricism from the intellectual domain of science! In this way, standard empiricism, as a result of controlling the flow of ideas and arguments in scientific journals, texts, lectures, degree courses and so on, effectively ensures that criticisms of standard empiricism of the kind developed here, however valid, are excluded from scientific discussion. Again, according to the philosophy of knowledge, more generally, essentially only claims to knowledge, and criticisms of such claims, may enter the intellectual domain of academic discussion. But what I put forward in this book is a criticism of the intellectual enterprise of giving intellectual priority to the search for knowledge and I put forward a counter proposal, namely that intellectual priority be given to realizing what is of value in life, and to proposing and criticizing possible actions designed to help achieve this. These arguments and counter proposals are not even intended to be contributions to knowledge: they are thus of a type that the philosophy of knowledge will exclude from academic inquiry.

The second point that needs to be taken into account is this. In seeking to assess the relative merits of the two philosophies of knowledge and wisdom we seek to decide an issue of immense and general importance: what the overall aims and methods of organized inquiry are to be, and how these are to be related to the aims and methods of life. We seek to decide nothing less than the question of how in general humanity should seek to think about and tackle its common problems.

All this serves to underline the importance of assessing the arguments for and against the rival philosophies of inquiry with the very greatest care and thoroughness.

As the above outline indicates, the central argument of this book can be interpreted as solving Wootton's problem of how to apply scientific method to social problems, so that something comparable to the extraordinary technical progressive success of natural science and technology may also be achieved in personal and social life. At the same time the argument offers a simple and general explanation as to why there has been such a glaring discrepancy between the rapid technical progress of science and technology, and the only faltering progress of humanity towards a better world.

In order to solve Wootton's problem, what we need to do is to apply directly to our problems of living appropriately generalized rational methods that have proved to be so extraordinarily successful in solving problems of knowledge in science, so that we come to solve our problems of living in the characteristically progressive way in which problems of knowledge are solved in science. This is in essence the philosophy of wisdom. Wootton makes the disastrous mistake of attempting to apply scientific method not to problems of living, but to problems of social knowledge. Rational methodology is applied not to social life but to social science. This is the basic mistake of the philosophy of knowledge. It is the long-standing persistence of this mistake which in part at least explains the glaring discrepancy between humanitarian and scientific progress. For this glaring discrepancy is due precisely to our long-standing failure to resolve in cooperatively rational ways our problems of living in the kind of way in which problems of knowledge are at present resolved in science.

It was of course the great hope of the 'philosophes' of the Enlightenment – the great hope of men like Voltaire, Diderot and Condorcet – that *scientific* progress might contain the key, the vital clue, to how humanity might achieve personal and social progress towards enlightenment (Gay, 1973). Indeed, the programme of learning from the progress of science how to achieve progress in life towards the realization of value may well be called *the Enlightenment programme*.

In order to implement this programme, however, two vital preliminary problems must be solved. First of all, a correct characterization must be given of the methodology that is actually in practice exploited by science and responsible for scientific progress. Second, this methodology must be appropriately generalized so that it becomes fruitfully applicable to all human endeavours, and not just the one endeavour of improving knowledge. In seeking to realize what is of value to us in life there are many goals that we seek to realize besides knowledge – such as health, happiness, friendship, love, justice, cooperative creative work, and so on. The crucial generalization, then, that needs to be made to scientific methodology so that it becomes a universally applicable progress-achieving methodology is to generalize the *aim* of the methodology, from the growth of *knowledge* to the growth of *value in life* in general.

The way to solve these two preliminary problems is spelled out in chapters 3, 4, 5 and 9. In terms of the argument of chapters 3 and 4, the progress-achieving methodology of science amounts in essence to (a) articulating, and trying to improve the articulation of, the basic problems to be solved, and (b) proposing imaginatively and assessing critically possible solutions. The way in which the philosophy-of-knowledge version of this progress-achieving methodology needs to be modified and generalized so that it becomes fruitfully applicable to all that we do (including science) was spelled out in chapter 4. In terms of the argument of chapters 5 and 9, the progress-achieving methodology in fact exploited in science with such astonishing success (even though this has rarely been understood) is *aim-oriented empiricism*. In order to become fruitfully applicable to all that we do, this needs to be generalized to become *aim-oriented rationalism*. The basic task of the Enlightenment programme is to help us cooperatively exploit problem-solving rationality, or aim-oriented rationality, in all that we do. This is the task of the diverse branches of social inquiry. As a result of implementing this Englightenment programme, we may well expect to achieve in life a degree of progress towards what is of value to us that is comparable to the remarkable progress that has been achieved in science (in improving knowledge). In seeking to make progress towards a better life the vital lesson to be learned from science is a *methodological* lesson. What science has discovered about the world is of course important: but the *manner* in which these discoveries have been made is perhaps of even greater importance. The cooperative rational progress achieved in science at its best has much of value to teach us about how to achieve cooperative rational progress in personal and social life.

The 'philosophes' of the Enlightenment might have succeeded in clearly articulating and advocating this programme in the eighteenth century. If they had, the unprecedented seventeenth- and eighteenth-century stream of scientific progress might well have burst its banks and flowed into all human endeavours throughout the eighteenth, nineteenth and twentieth centuries, thus becoming an unprecedented flood of social, humanitarian and spiritual progress. What began as a rapid growth of knowledge would have broadened into a rapid growth of social wisdom throughout the world. The present-day discrepancy between scientific success and human failure would not exist.

But this did not happen. The 'philosophes' of the eighteenth century disastrously misunderstood the Enlightenment programme. Instead of endeavouring to apply aim-oriented rationalism to personal and social life, thus developing social inquiry as social methodology, they sought to apply scientific method to the task of developing *social science*. And as we have seen in chapter 6, this disastrous perversion of the Enlightenment programme has persisted down to the present day. Thus, the human disasters of the twentieth century are due to our failure to put right an intellectual disaster of the eighteenth century.

Is this conclusion correct? Granted that we have indeed failed to put right a disastrous intellectual mistake of the eighteenth century, would correcting this mistake really have such extraordinary consequences?

At least three reasons can be given as to why it is more or less inevitable that nothing comparable to the rapid, accelerating technical progress of modern science and technology can be introduced into the rest of human life, to enable us all to make rapid human progress towards a more just, free, civilized and loving world.

First, for rapid scientific and technological progress to be made it is not necessary that we all take part: it is only necessary that relatively few, highly talented and motivated people be trained in scientific and technological research, and be given the opportunity, by funding and so on, to take part in such research. By contrast, if we are to make real human progress towards a better, more humane, more civilized world, then we all (or almost all) need to take part, the intelligent and highly motivated, and the stupid, the unmotivated, the power-mad, the careerists, the manipulative, the criminal, the dispossessed, the hopeless, ill, mad and dying. We cannot expect a relatively small army of paid experts to solve the world's problems for us in the same way as we may expect such an army to solve for us our scientific and technological problems.

Second, natural science and technology have the immense advantage of being able to employ the method of experimentation, of relatively uncostly trials. In these fields, we can perform experiments, and build and test material models, in order to try out our scientific and technological ideas, without our having to pay the price of widespread suffering, injury and death when our trials fail (assuming reasonable precautions are taken). There are of course limits to what we can do, and ought to seek to do, even in the domains of the physical and technological sciences, let alone the biological and medical sciences. Nevertheless, freedom to try ideas out painlessly is immense in comparison to what is possible for us in human affairs – in politics, education, industry, the media, commerce, international relations, bringing up children, our own personal lives. Here, inevitably, people are involved: to experiment, to try out possibilities here is to experiment with people's lives. Failure is not just six months' paid work down the drain: it may involve appalling suffering, wasted lives, and may be irreversible in that human experiments, once started, may be impossible to stop even if obviously undesirable.

Third, in science and technology failure is often obvious and uncontroversial: theories are constantly refuted experimentally, and prototypes for new technology can often readily be seen to be failing to work as expected. What counts as success and failure is relatively unproblematic. In our personal and social lives, success and failure is rarely as obvious and unproblematic. Even if, in real terms – in terms of our original aspirations, or in terms of what is of real value to us – we are failing miserably, nevertheless we may all too successfully conceal from ourselves the fact of our failure – or, alternatively, in connection with political or institutional action, failure may be ignored by those with power. And not only is it more difficult to detect failure in human affairs than in science and technology: in addition what is to count as success and failure is much more problematic, and differs from person to person, and from group to group.

In short, we cannot reasonably expect to be able to learn from our mistakes in life in anything like the rapid, progressive way in which we learn from mistakes in science and technology – partly because in life we cannot hire clever experts to do our thinking for us, partly because in life we cannot deliberately and painlessly make lots of mistakes from which to learn, and partly because in life mistakes are often difficult to detect and agree about.

However, the fact that learning and progress in human life are inevitably more difficult and problematic than narrow intellectual and technical learning and progress in science and technology, as these three considerations indicate, does not provide us with any good reason for not attempting to introduce into life the progress-achieving methodologies already so successfully exploited in science. It is all the other way round. Just because especially severe difficulties do arise in putting into practice progress-achieving methodologies in life, the whole endeavour deserves all our attention and care. Thus, for example, sustained attention needs to be given to the multitude of problems generated by the following basic question: How can we build into our political life and institutions methodological principles designed to help us discover and achieve generally desirable personal and social or political objectives, such as freedom, justice, prosperity, opportunities for cooperative work and endeavour, the capacity to resolve conflicts in just and mutually beneficial ways? And quite generally, sustained attention needs to be given to the multitude of problems that arise in connection with analogous questions to be asked about all the other aspects of our personal and communal lives. In doing this we must of course take into account that we are a mixture of the dedicated and idealistic, the ambitious and unscrupulous, the intelligent and stupid, noble and criminal, rich and poor, well and ill. It will be especially important to develop education for everybody as discovering how to put into practice and develop progress-achieving methodologies in life, in diverse personal and inter-personal pursuits, so that we may realize what is of value to us. Just because we cannot experiment in personal and political life in the free way in which we can in scientific and technological research, it becomes all the more important that we learn all we can from the variety of actions people do perform, and have performed, in attempting to resolve problems of living; and it becomes all the more important that we create vividly and accurately imagined trials, possible deeds, so that we may learn from what we imagine ourselves doing and do not do, and not only from what we actually do. Just because what is of value in life is problematic, and different for different people and different groups of people, success and failure in life being problematic and diverse, all the more do we need to attend, imaginatively and critically, to questions about what kind of success we really do want to achieve, what kind of failure we want to avoid. And just because failure is often difficult to recognize in life – painful to acknowledge – we need to give sustained attention to the task of developing traditions and habits that help us to recognize and acknowledge failure when it happens (most of the time!).

In brief, we only have a reasonable chance of successfully and progressively realizing what is of value in all the diverse pursuits and aspects of life if we inherit and can make use of a tradition of organized inquiry and education that gives absolute priority to the tasks of developing and helping us to put into practice method-ologies designed to enable us to achieve such success and progress in life.

At this point it may be conceded that having a tradition of rational inquiry devoted to the growth of wisdom is a *necessary* condition for developing a wiser world; and yet the importance of trying to develop such a tradition of inquiry within research and educational institutions may nevertheless be denied. In support of this denial, the following arguments may be produced: (1) social factors external to universities and schools would inevitably make it impossible to put the philosophy of wisdom into practice: it would never be permitted by governments, public opinion, religious authorities; (2) internal factors would make it impossible to put the philosophy of wisdom into practice: scientists, scholars, teachers, administrators would never be able to agree sufficiently amongst themselves about political, moral, social, religious or ideological issues for a good enough consensus to develop to make cooperative, rational, intellectually productive exploration of personal and social problems of living a possibility; (3) even if it did prove possible to put the philosophy of wisdom into practice intelligently and fruitfully in universities and schools, this would nevertheless have only a negligible influence on the rest of the world, as good proposals for action emerging from universities would be systematically ignored; (4) there is no real need to put the philosophy of wisdom into practice in universities since there are no substantial intellectual problems about what we need to do in order to solve our major social problems. What we need to do is obvious; the problem is the political one of persuading others of the need to do the obvious (sometimes against their own short-term interests).

My reply to these arguments is as follows. (1) In many places in the world political and religious power may well at present make it impossible to put the philosophy of wisdom into practice in universities and schools. The extent to which the thing can be done in secular, democratic nations can only be discovered by trying. (2) Granted that external pressures do not prohibit putting the philosophy of wisdom into practice, it may well be difficult to develop and sustain traditions of tolerance, of cooperative, open and critical discussion, of learning from opponents, about potent political, moral and ideological issues, within universities and schools, so essential for putting the philosophy of wisdom into practice. It is not always easy to imagine schools and universities in which Marxists and Tories discuss political issues together, intelligently and with a common good will. It is all too easy to imagine schools and universities being taken over by some one ideological, political or religious doctrine, or by some one powerful group with its own special interests, by means of an appropriate policy of appointments and redundancies, operating behind a mask, perhaps, of 'open, critical, pluralistic inquiry'. The internal difficulties confronting the task of developing a genuinely rational kind of inquiry (devoted to the growth of wisdom) are considerable even in democratic societies, but not, I maintain, insurmountable. (3) It is entirely proper that universities should have influence but not power (otherwise one has a modern version of Plato's Republic). Inevitably many valuable proposals and criticisms emerging from universities (devoting reason to the growth of wisdom) will be ignored or rejected by society, by the political and economic system (an inevitable consequence of lack of power). Nevertheless, the existence of a vocal, active tradition of rational discussion of social problems can still profoundly influence thought, policy and action in the broader social world – via education, lectures, books, articles available and understandable to non-academics, journalism, the multitude of formal and informal points of contact that exist between universities and society. The activity of articulating and exploring proposals for action in public can in itself help make possible social action that would otherwise be impossible. A society in which there is a tradition of rational discussion of its problems has open to it all sorts of desirable possibilities – in particular the possibility of democratic, non-authoritarian, cooperative action – not open to a society in which there is no such tradition. (4) It is admittedly often held that no problems arise as to what needs to be done in order to solve our social problems, realize what is of value in life, all the problems having to do with persuading or forcing others to act appropriately. Amongst those who hold this view, however, one finds an incredible diversity of views about what does need to be done in order to resolve our social problems. When one takes into account the immense complexity of these problems, the almost inevitable capacity of social action to have all sorts of unforeseen consequences, the immense diversity of character, circumstances, capacities, aspirations, desires and fears of people in society, it is difficult not to regard the idea that social problems have obvious solutions as utterly idiotic, an absurdity scarcely worth mentioning were it not for the fact that it is such a widely held and dangerous illusion. Perhaps one should rather say that any adequate solution to a social problem, requiring many people to act cooperatively, must almost inevitably be of such complexity, requiring such a diversity of actions from those who participate, that no one person can hope to have anything but the vaguest notion of what the 'solution', the 'action' amounts to. Human life is so rich and diverse that even our own experiences and actions are beyond our full comprehension, let alone those of many people taken together. One important initial contribution that inquiry, pursued in accordance with the philosophy of wisdom, would be in a position to make, would be just to render commonplace the Socratic idea that we are all more or less ignorant of what is of most value in life and of how it is to be realized, learning in this domain being both possible and supremely desirable (an idea that is at present almost a commonplace within science with respect to knowledge).

In order to highlight the difference in *method* advocated by the philosophies of knowledge and wisdom for social inquiry, let us consider briefly the contrasting approaches to the following somewhat humdrum social problem: the progressive dereliction of city centres.

A social scientist taking the philosophy of knowledge for granted, but anxious nevertheless to make some kind of contribution to our problem, will proceed more or less along the following lines. Armed with some kind of provisional understanding of the social problem in question (without which one would not be in a position to proceed at all) he will seek to gather social data which he deems to be in some way relevant to the problem. If he is relatively sophisticated methodologically, he may well put forward a conjecture designed to explain the progressive dereliction of city centres, which he then proceeds to attempt to refute or corroborate empirically. He may even attempt to gather data designed to decide between two conflicting theories. He may make a comparative study of two cities, one of which only is deemed to be progressively decaying. In any case, he will gather data, of a statistical character, by carrying out a survey, getting a randomly chosen sample of the population to fill in carefully prepared questionnaires. He will arrive at certain empirical conclusions, which may include a corroborated theoretical explanation of the social, cultural, economic, legislative factors which cause decay of city centres, but he will not come up with a proposal as to how the problem can be solved. His task is to solve a sociological problem – a problem of knowledge and understanding – not a social problem.

Apart from the obvious criticism that such a social scientist does not even attempt to help solve the basic social problem, there is a further serious criticism to be made. The whole way in which the social scientist proceeds, the data he seeks to gather, the empirical theories he is prepared to consider, will be profoundly affected by his initial understanding of the underlying social problem, the way he formulates it, the kind of policy-measures he is prepared to consider as reasonable. And yet, just because such a social scientist seeks to solve a *sociological* problem, not a social problem, no explicit analyses or discussion of the underlying social problem is likely to be given. Implicit assumptions concerning human, social priorities and political options may well profoundly affect the kind of data that the social scientist seeks to gather: and yet these assumptions will not be explicitly articulated and critically assessed. As a result, the data that the social scientist gathers may well only be of interest to those who agree with the way the underlying social problem has been understood. Even worse, presenting the data as objective, value-neutral, politically-neutral, empirical results, when in fact value-judgements and political judgements are bound to be implicit in decisions as to what sort of data are significant and relevant, will have the effect of influencing the reader of the eventual report to accept uncritically the underlying understanding of the social, political problem. Implicit, covert presuppositions are always much harder to challenge and resist than explicit assumptions. Thus the social scientist's work has the effect of obstructing the one thing that ought to be promoted – explicit, critical articulation and analysis of the underlying social, political problem.

Finally, of course, the social scientist's results can only aid manipulative social, political action. As a result of discovering that people are influenced to act in such and such ways by such and such factors, one basis is provided for enacting new legislation, for example, designed to influence the people involved to act differently.

A social inquirer who approaches the problem from the perspective of the philosophy of wisdom will proceed in an entirely different way. His approach will be much more like that of a good journalist than that of an orthodox academic social scientist.[1] His basic task is not to improve empirical knowledge of social phenomena at all:[2] rather it is to engage in, and to help promote, rational approaches to solving the basic social problem.

It is important to recognize that a major problem that confronts any attempt to resolve the social problem in a cooperatively rational fashion arises simply from the number of people that are involved. If an analogous problem confronted a tribal village, in that the centre of the village was suffering from progressive decay, it would always be possible to hold a meeting, which all members of the village could attend, at which the problem could be discussed, and an agreed policy be decided upon. Modern cities cannot cope with their problems in an analogous fashion. An important long-term problem is precisely to develop institutional machinery which enables us, as far as possible, to overcome this obstacle to rational, cooperative social problem solving.

In the absence of a solution to this long-term problem, our social inquirer must do the best that he can. His task will be to enlist the

---

[1]It is my impression that much of the best work produced in the field of social inquiry is indeed produced by writers who proceed as good journalists – such as Daniel S. Greenberg, Anthony Sampson, Tony Parker, Richard Barnet, Ronald Blythe.

[2]I am not, of course, arguing that social inquiry should not seek to acquire factual knowledge at all. Rather, I am arguing that factual knowledge should be sought as a secondary, subordinate intellectual task, both within social inquiry, and within inquiry as a whole, subordinate to the fundamental intellectual tasks of articulating problems of living, proposing and criticizing possible solutions, possible actions designed to help us realize what is of value in life. In doing this, we make use of, and improve as needed, our already possessed highly sophisticated knowledge and understanding of ourselves, each other, institutions, social structure, the material world. As long as intellectual priority is given to the promotion of rational, cooperative problem-solving, the development of predictive knowledge of human behaviour is not harmful, and may well be beneficial – insofar as such knowledge is used to promote cooperative problem-solving in life.

help of some of those involved in the problem, in one way or another, in an attempt to improve his understanding of the problem, and improve his ideas as to what policies might be developed which, if put into practice, would help solve the problem. His concern will be to provoke people into putting forward suggestions, proposals, and into criticizing the suggestions and proposals of others. In addition he will himself discuss and criticize the ideas of the people he interviews. It will be essential for him to interview people involved in the problem in different ways: house owners, flat dwellers, property developers, business-men, government officials, politicians, social workers, the police, pressure groups, shopkeepers. His task will be to probe beneath rhetoric to underlying aims, actions and motivations. And finally, in writing up his report, his concern will be to leave a record of his attempt to find a possible solution to the problem, publishing the ideas, arguments and responses of those he interviews, as well as an account of their actions. He will be concerned to make his report as clear, and interesting to read as possible, and have it published in a generally available form, so that it in turn may stimulate more enlightened public discussion of the issues involved.

The result of the failure of social inquiry to give intellectual priority to the task of promoting and sustaining cooperative, rational, problem-solving in the world is of course that such problem-solving fails lamentably to flourish. Consider, for example, the extent to which we succeed at present in tackling our economic problems in a cooperatively rational fashion.

At present the economic system that prevails in free market democracies is such that most adults receive treatment appropriate to children. There is, for most, no opportunity to take part in a jointly-owned, cooperative venture, where ownership, manage-ment, risk and responsibility are shared by all those who work for the venture. Instead, most people work for a wage, as instructed, without responsibility, part-ownership, or managerial influence, for the profit of the employer or share-holders. If indifferent work, unrealistic wage claims and strikes result, this should occasion no surprise. Treated by the system as an irresponsible child, it is scarcely surprising if one responds in kind.

One exception to all this is the extraordinarily successful cooperative movement of Mondragon in Spain. And the striking fact about this movement is that it began with some practical (aim-oriented rationalistic, philosophy of wisdom) economic inquiry initiated by a priest, Jose Maria Arizmendi. As a result of his own research into earlier cooperative movements – such as that of

CHAPTER EIGHT

Objections to the Philosophy of Wisdom

Despite the arguments of the previous chapters, there may be some who wish to hold on to the view that the basic intellectual aim of inquiry should be to improve knowledge and not wisdom. The following arguments may be given in support of the philosophy of knowledge and against the philosophy of wisdom.

1   All rational inquiry, all rational thought has, as its aim, to establish or improve knowledge. Even in the context of action, the rational component of thought is devoted to establishing knowledge of the truth or falsity of various conditional propositions such as 'if X is performed, Y will result' or 'in order to realize A in the easiest and quickest way, B must be performed'.

2   Some reasons have previously been given for holding that, in order to give science a good chance of serving humanity, discussion concerning the aims and priorities of research, and the social use of results of research, need to be rationally related to discussion of social problems of living. This does not however in any way undermine the central tenet of standard empiricism and the philosophy of knowledge which asserts only that the *results* of research, namely claims to knowledge, must be assessed with respect to truth alone, independently of all consideration of aims of science and of life, human needs, desires, feelings, objectives, values.

3   The scientific pursuit of knowledge may be undertaken merely in order to acquire knowledge, and not at all in order to help promote human welfare. In this case, all the arguments against divorcing scientific problems from social problems – against standard empiricism and the philosophy of knowledge – become irrelevant.

4   The philosophy of knowledge is only vulnerable to criticism because it has been put forward in a grotesquely inflated form as a theory of *all* of rational inquiry. Reinterpreted more modestly, as a theory only of that part of inquiry that is devoted to the acquisition of knowledge, it becomes entirely acceptable, and immune to the criticisms that have been levelled against it.

5   It is essential that universities restrict themselves to devoting inquiry to the acquisition and improvement of knowledge. By doing this, universities can perform a vital service to the community, and indeed to humanity, while at the same time having some chance of retaining their intellectual independence from government. The moment universities adopt as their official intellectual aim to help develop a more rational world they must lay themselves open to charges of political, religious, moral and ideological bias, given the wide range of interpretations that the idea of a 'rational society' is open to. Universities risk becoming dominated by government, and may well become the slaves of political, religious, moral or ideological dogma, or the servants of those with power and money in the community; or they may become a battleground of sterile controversy between rival factions within universities, the vital, more modest task of improving knowledge being neglected. The enterprise of helping to develop a more rational, cooperative and humane world is a vital one: but it is a political or moral enterprise, which needs to be pursued in the world, outside the groves of academe. The enterprise of helping to develop a more rational, a wiser world, conducted within universities, can have little impact on the world itself, and only a destructive impact on proper university activities having to do with scientific research, scholarship and education.

6   Acquisition of relevant knowledge is an essential prerequisite for the rational tackling of problems of action, problems of living. Far from action, and problems of action, being intellectually more fundamental than knowledge and problems of knowledge, as the philosophy of wisdom maintains, it is all the other way round. Acquisition of knowledge is intellectually prior to and more fundamental than the rational tackling of problems of action. Without knowledge, action becomes impossible. Without knowledge it even becomes impossible to imagine possible actions, and thus to engage in the rational tackling of problems of action. Thus, the central assumption – the whole basis – of the philosophy of wisdom is untenable, and is to be rejected.

7 The philosophy of wisdom – with its emphasis on solving practical problems of living – cannot do justice to the intellectual value of inquiry, the value of inquiry pursued for its own sake without ulterior motive. Only the philosophy of knowledge can do justice to this aspect of inquiry.

8 The philosophy of wisdom advocates a species of 'Utopian social engineering' – a kind of social planning that has been decisively criticized by Popper.

9 The philosophy of wisdom, in committing inquiry to *cooperative* social problem-solving, commits inquiry to a programme of action that constitutes a massive infringement of individual liberty. The philosophy of wisdom is thus to be opposed in order to protect individual liberty from being drowned in an ocean of 'cooperativeness'.

My reply to these counter doctrines and arguments is as follows:

Reply to objection 1
In order to tackle our problems of living rationally, we need at the very least, according to the philosophy of wisdom, to engage in the intellectual activity of imagining and criticizing possible actions from the standpoint of their capacity to solve our problems. This involves, but is certainly not equivalent to, considering propositions of the form 'if X is performed, Y will result'. The intellectual excellence of our thinking is to be judged in terms of how good our imagined actions are, and how well assessed, from the standpoint of their capacity, if enacted, to solve our problems in such a way that we realize what is of value to us. For intellectual excellence, it is essential that propositions considered of the form 'if X is performed, Y will result' are not only *true*, but are also *relevant*, conducive to a good resolution of the problem in hand. Thus, according to the philosophy of wisdom, for inquiry to be rational, it is vital that it is not reduced merely to the consideration of claims to knowledge of the type 'if X is performed, Y will result'. To this one might add that in order to be rational, the intellectual activity of imaginatively exploring possible actions needs to be linked to and motivated by the desire and the capacity to *act*, when a good action is discovered: otherwise cogitation will be in vain. Rationality demands, in other words, that thought and action be interlinked in a certain way, whereas of course the philosophy-of-knowledge conception of reason demands that inquiry be divorced from action if inquiry is to be rational!

Reply to objections 2, 3 and 4

These objections claim, in various ways, that standard empiricism and the philosophy of knowledge are acceptable as long as they are interpreted sufficiently modestly as doctrines about how to acquire knowledge only, with no import as to how knowledge is to be applied so as to promote human welfare. At least three arguments have, however, already been developed against such 'modest' versions and defences of the philosophy of knowledge. (a) The philosophy of knowledge misrepresents the basic intellectual aim of science, in that it fails to do justice to the search for *understanding*. (b) In addition, it fails to do justice to the aim of improving knowledge of *valuable truth*. Values inevitably, and quite properly, exercise a major influence over the intellectual domain of science, over estimations of scientific progress. Standard empiricism and the philosophy of knowledge, in banishing discussion of untestable ideas, and conjectures and problems about what is of value, from the intellectual domain of science, serve to undermine the intellectual rigour and success of science. (c) Whatever else it may be, science is an expensive and influential human enterprise: it needs therefore to be assessed as a human enterprise, a social or institutional activity with typical social or institutional aims, achievements and problems. In insisting that science be conceived and assessed in exclusively intellectual terms, the philosophy of knowledge illegitimately deflects valid criticism, of a social and moral character, away from science. (Arguments against an even more 'modest' version of the philosophy of knowledge are developed in the next chapter.)

Reply to objection 5

As long as the enterprise of improving knowledge can be conducted rationally when intellectually and institutionally dissoci-ated from rational discussion of social problems of living, a defence of the view that universities ought to restrict themselves to improving knowledge is at least possible. But the arguments of chapters 3–5 have shown that the pursuit of knowledge must in important respects cease to be rational, if dissociated from concern with social problems. Hence the above view becomes indefensible. It may well be that attempting to put the philosophy of wisdom into practice in univesities faces greater dangers and difficulties than putting into practice the philosophy of knowledge. This does not in itself constitute sufficient grounds for not making the attempt. In the past men like Bruno, Copernicus, Kepler, Galileo, Descartes might well have decided that the dangers and difficulties associated with the rational pursuit of knowledge of nature were too great for the thing to be attempted: and as a result, no doubt, modern science as it is today would not have come to be.

Reply to objection 6
The argument that the acquisition of relevant knowledge must precede, and be intellectually more fundamental than, rational action is perhaps the central argument in support of the philosophy of knowledge. Widespread conviction in the validity of this argument is perhaps responsible, more than any other intellectual factor, for the persistent domination of the philosophy of knowledge in academic institutions. I therefore now devote some space to demonstrating that this argument is invalid, and that, quite to the contrary, the philosophy of wisdom is absolutely correct in insisting that more or less successful action in the world, and rational tackling of problems of action, are prior to knowledge. I have four main points to make.

1   Even if the objection were valid, this would still not undermine a central contention of this book, namely that rational inquiry devoted to promoting human welfare must devote much attention to articulating problems of living, proposing and criticizing possible solutions, this being intellectually integrated with science, the pursuit of knowledge.

2   The validity of the objection becomes extremely doubtful when one reflects on the extraordinary extent to which practical problems have been successfully solved in the past in a state of extreme ignorance – and when one reflects on the inevitability of ignorance. Endlessly many examples can be cited – from social life, from technology, and from medicine – of problems of action being successfully solved in the absence of what can only be regarded as relevant knowledge. Much of the basic technology possessed by mankind, upon which the modern world is founded, was developed by primitive man in prehistory, long before adequate theoretical knowledge and understanding of the relevant phenomena had been developed. Hunting, agriculture, irrigation, husbandry, fire, cooking, smelting, metalwork, pottery, clothing, building, transport, medicine: basic discoveries in all these fields were made by people who were convinced that the natural world is animated by gods, and who thus severely lacked relevant knowledge. And indeed, not only does technology come before science: the subsequent development of science is scarcely

conceivable without the prior development of some basic technology. Even after much scientific knowledge has been accumulated, technological discoveries have continued to be made before adequate relevant theoretical knowledge and understanding has been achieved, especially perhaps in medicine. And even today we have every reason to believe we are still profoundly ignorant of the nature of the ultimate constituents of the world, of the fundamental laws of nature, and of how our brains function: in a sense our whole life is conducted within an ocean of ignorance about the world, ourselves, our immediate environment. Our capacity to acquire relevant knowledge before we act must inevitably be severely limited – and even the process of acquiring knowledge itself requires that we act in a state of ignorance.

3   This objection is not valid. In order for rational tackling of life problems to become possible it is not knowledge that we need so much as *conjectures*, and the capacity to learn. We can then acquire knowledge about our environment and ourselves as needed, as we tackle our problems of living. Proposing and criticizing possible actions in the absence of knowledge *is* possible, and is in fact essential for rationality. Indeed, in practice, we are almost all the time obliged to tackle our life problems in a profoundly ignorant state about all sorts of matters of possible relevance to our actions – for example, the intentions, the future actions, of other people. It would actually be appallingly irrational to attempt to acquire knowledge of all relevant factors before tackling problems of action – simply because the endeavour would be never-ending, the prescription thus leading to complete paralysis. Indeed, if Popper is correct in holding that, strictly, we cannot have knowledge at all, but at best only well-tested conjectures, to wait for knowledge before one is prepared to act is simply to cease to act forever.

This argument can be put in a particularly forceful way as follows. There is scarcely any part of our environment which exercises such a profound influence over our conduct as our own brains. Therefore, if we take seriously the principle that knowledge of relevant factors must first be acquired before rational action becomes possible, we must first acquire knowledge of our brains before we act (if we are to be rational). Such advice is clearly absolutely disastrous, since we still do not really know how brains work even in broad outline, and probably cannot ever know, even in principle, for logical reasons, what is going on in detail in our own brains. Furthermore, if we were to take the advice seriously we would actually forego forever the possibility of improving our neurological knowledge, since neurology, like the rest of science, can only be pursued if we can act successfully in the world – a strand of the argument to be developed below.

But the argument needs to be developed even more forcefully than this. It is not just that in practical contexts rational action always proceeds in a state of enduring ignorance. Even when our prime concern is to improve knowledge, to do science, nevertheless priority still ought to be given to practical problems of action, to questions of what we want to do, what we want to achieve: for only in this case can we be in a position to know what new knowledge and technology it is relevant for us to try to develop, in order to make possible, or reveal the impossibility or undesirability of, proposed actions.

The argument here is essentially analogous to Popper's argument in support of hypothetico-deductivism as opposed to that version of inductivism which holds that evidence must first be accumulated before sound scientific theorizing can begin. Popper argues, in effect, that it is only if we give intellectual priority to articulating and criticizing *theories* in the natural sciences that we can know what observations and experiments it is relevant to make in order to test our theories. Accumulating evidence without prior theorizing leads only to a mass of trivial, useless results (Popper, 1959, pp. 106–8). My point is essentially analogous to this. It is only if we begin with proposals for action that we can know what scientific knowledge and technology it is relevant to try to develop in order to assess critically, or implement, these proposals. The pursuit of knowledge and technology without prior critical thought about what it is that we want to achieve is likely to lead to a mass of trivial, useless results when judged from the standpoint of achieving what is most humanly desirable. Both points are applications of the elementary point that rationality demands giving absolute intellectual priority to articulating and criticizing possible solutions to the problems to be solved. Popper, concerned primarily with inquiry devoted to solving theoretical problems in the natural sciences, stresses the intellectual priority of proposing and criticizing *theories* – possible solutions to these problems. I, concerned primarily with inquiry devoted to helping to solve problems of living, stress the intellectual priority of proposing and criticizing possible and actual *actions* – possible solutions to these problems. Consideration of our life problems, in short, constitutes a proper rational spur for the development of relevant knowledge and technology.

4   But in addition to this, the argument (that knowledge must first be possessed before rational action becomes possible) is wrong in an even more radical way – if by 'knowledge' is meant explicit propositional knowledge, rather than merely the capacity to act successfully in the world. In fact, I wish to argue, the thing is all the other way round. It is only insofar as we can act successfully in the world, and can propose and assess possible actions, that we can be in a position to possess or acquire explicit propositional knowledge. Action, and the ability to imagine oneself performing possible actions, must first come into the world before there can be explicit propositional knowledge. 'Knowledge', in this sense, is a development of, and not a prerequisite for, rational action.

I suggest the evolution of thought and knowledge in the world need to be conceived of in the following terms. To begin with there develops the capacity to act successfully in the world, the capacity to realize life-goals, solve problems of action. Fish, beetles, ants and spiders must be able to eat, to avoid being eaten, and to fertilize eggs or mate, in order to reproduce. The ability to reproduce depends crucially and fundamentally on the ability to realize successfully such goals, solve successfully such problems of action. Natural selection operates primarily on this ability. But at this primitive level there is no conscious thought or knowledge. In so far as primitive life forms can be said to possess implicit knowledge of the environment and to possess the capacity to acquire such knowledge, this must be understood in terms of the ability to act, to realize life-goals, to solve in practice problems of action.

At the most primitive level, an organism's repertoire of possible actions is largely genetically determined. However, even at the most primitive level, there must be some flexibility in action, in that action is a product of genetic determination and environment, and the environment varies with time and place. At a somewhat higher level there is learning, a flexible rather than fixed repertoire of actions plus the ability to learn being genetically determined. Animal learning is, however, essentially learning how to act more successfully. Even perceptual learning about the environment – which seems somewhat like the acquisition of knowledge – must be understood as an aspect of action, an aspect of solving problems of action. Enhanced powers of perception have survival value insofar as they enhance the ability to act successfully. At a still higher level there is imitative, social or cultural learning rather than only learning at the individual level: lion cubs learn hunting skills through play, through practice on half dead animals caught by parents for that purpose, and through imitating adult lions while hunting. The schools and academies of Nature are firmly based on the assumption that the purpose of education is to learn how to act, to live, to solve problems of living.

At a still more sophisticated level, there develops the capacity to *imagine* that actions are being performed. A lion hunting a zebra, for example, imagines various possible lines of action: various possible routes to the zebra are rehearsed in the imagination. Consider two apes confronted by the kind of problem with which Köhler once used to plague apes: bananas hanging out of reach can only be obtained by piling three boxes on top of each other. The first ape, let us suppose, tackles the problem by trying out all sorts of more or less unsuccessful actions until eventually the problem is solved. The second ape, at the other extreme, tries out all sorts of more or less unsuccessful actions *in its imagination* until the problem is eventually solved, and the ape swiftly puts the solution into practice. This second way of solving the problem seems on the face of it much more impressive as an intellectual performance than the first way: we may be misled, like Köhler and others, into invoking some mysterious mental act of insight to explain the apparently sudden emergence of the completed solution. The greater apparent intelligence or rationality of the second ape is, however, in a sense deceptive: it is due simply to the fact that the second ape blunders about stupidly in the privacy of its imagination, thus rendering its nearly random, stupid efforts to solve the problem invisible to our eyes. There is thus no need to appeal to 'insight' here – or rather 'insight' can be fully explained by the assumption that the second ape is able to learn from what it does in its imagination while the first ape is restricted to learning from what it does in reality.[1]

The extent to which animals are in fact capable of this kind of imaginative problem-solving is at present no doubt a controversial issue. At this point I wish only to stress the following points. Being able to try out possible solutions to problems of action in the imagination – as opposed to being restricted to trying them out in practice – clearly has great potential survival value, especially for hunting. Thus there is no problem in understanding why natural selection should favour the development of such an ability. The hypothesis that the higher animals are capable of imagining actions, at least to some extent, provides an explanation for the otherwise puzzling phenomenon of dreaming. The evolutionary

---

[1] See Köhler (1927) for his own account of insight in apes.

advantage of dreaming is generally conceded to be problematic. From the standpoint of the present hypothesis, however, dreaming may be understood as Nature's way of enabling an animal to develop its ability to imagine it is performing actions not actually being performed: this, after all, is the crucial feature of dreaming. This hypothesis thus explains why dreaming does have survival value. The fact that only the higher animals appear to dream supports the hypothesis – in that one would expect the ability to imagine actions to develop only with the higher animals.[2] Finally, the ability to imagine one is performing actions that one is not performing must be understood as a development of the prior ability to act. In imagining that it performs certain actions, an animal in effect arranges to have occur in its brain neurological processes that in certain relevant respects resemble those neurological processes that would occur in its brain were it actually performing the imagined actions. Both the potential survival value, and the meaning, of a process of imaginative thought, of the kind being considered here, require that imaginative thought be interpreted as the occurrence of inner processes analogous to inner processes that would be involved in the control of the imagined actions, were these actions actually to be performed. These inner processes are, we may legitimately conjecture, brain processes. To imagine, to think and to dream is to act with action suspended: what is being done makes essential reference to action, and to the prior capacity for action. Thus the ability to imagine, to think, must be understood as a development of the ability to act. (The theory of imagination just outlined might be called the 'suspended action' theory of imagination, of thought.)

At the neurological level, the ability to imagine action can perhaps be understood as a development of the particular way in which action is neurologically programmed and controlled in order to facilitate learning.

Complicated actions performed by a spider, for example, in making a web, are, we may suppose, specified and controlled neurologically step by step. Nothing prompts the spider to engage in *general* web-making behaviour, and nor is the outcome dependent on prior web-making behaviour, on what has been learned. On the contrary, the spider is induced to perform a

---

[2]The theory of dreaming proposed here is diametrically opposed to Sagan's theory, according to which the dream-state involves the activation of primitive systems in the brain, inherited from our reptilian past, which are repressed when we are awake. See Sagan (1978) pp. 149–51.

specific sequence of actions which result in the construction of a web of a certain definite structure which is neurologically, and ultimately genetically, predetermined. Variations in webs built by spiders of the same species are to be explained in terms of variations of the environment, and the accidents of construction.

A kitten, however, is induced to engage in hunting behaviour in a quite different way: there is an impulse, a desire, to act out general hunting behaviour, the actual performance thus depending crucially on prior practice. From a neurological standpoint, we may suppose that something in the kitten's brain induces neuro-logical processes to occur which are somewhat analogous to those which occur when the kitten crouches and pounces. The kitten begins to imagine, to dream while waking, that it is hunting, and thus is induced, in a flexible way, to act out what is imagined. The imagination, on this view, is an internal arena within which desired actions are performed, thus inducing actual actions in a flexible fashion, in a fashion which facilitates learning. The capacity to imagine actions is, on this view, a development of action being neurologically controlled in a way which facilitates learning, by means of desire, motive or instinct which flexibly, rather than precisely, determine action.

Be this as it may, the point that I wish to stress is that the development of the ability to imagine action – itself a development of the prior ability to act successfully in the world – is the crucial step in the development of thought, understanding of others and oneself, self-awareness, rational problem-solving (as defined above), propositional knowledge and science.

Being able to imagine action makes possible an enormous extension in space and time of the context, the environment, in which an animal or person acts – in that possible actions at distant times and places can be imagined or conceived of. Long-term imaginative planning becomes possible – and also (of fundamental importance for the shaping of humanity, the shaping of the human predicament) realization of the inevitability of death. In addition, understanding of others becomes possible. If a person can imagine that he is in a context very different from his actual context, acting, seeking goals, having desires, feelings, experiences very different from his present actual actions, goals, desires etc., then, by extension of this, a person can imagine that he is another person, with that other person's goals, problems, context, desires, feelings, experiences. Imaginative understanding of others becomes poss-ible. It is this, I suggest, which leads to self-consciousness, self-awareness. As a result of imagining that we are other people, we become aware of ourselves as imagined by others. We see ourselves from outside ourselves, through the eyes of others: we become aware of discrepancies between the way others conceive of us, and our own experience of ourselves. It is this which produces self-consciousness, self-awareness. Self-consciousness is, on this view, essentially social in character, an outcome of seeing ourselves from the standpoint of others. Thus, our ability to plan, to understand others, and to be self-aware all arise out of the ability to imagine action, to dream while awake.

Finally, with the development of language, inter-personal or social imagining becomes possible. A group of people can cooperatively imagine, propose and assess, possible actions. And out of this develops the possibility of the cooperative development of shared knowledge and understanding – myth, common-sense, literature, propositional knowledge and science. Factual propositions about the world are, as it were, truncated proposals for action, truncated imagined actions, with the imagining, acting subject ignored or suppressed. According to this view, acquiring scientific knowledge about the world, and acquiring empathetic, imaginative understanding of others, are but two sides of the same coin, each dependent on the other. To imagine oneself to be another person successfully demands, amongst other things, the ability to imagine that the world is as experienced and conceived of by the other person. Just this imaginative entering into alternative possible worlds is at the heart of the natural sciences: in understanding rival physical or cosmological theories we in effect imagine ourselves to be people who conceive of the world in terms of these theories; what we do is a development of the kind of empathetic, imaginative understanding of others, which we all, to a greater or lesser extent, practise in life, and which is practised in a professional way by good psychiatrists, teachers, biographers, historians, anthropologists, actors and novelists. Understanding physical theory, and understanding people, are both, equally, developments of the basic ability to imagine oneself to be performing diverse possible actions.

It may be objected that the theory of the evolution of thought, consciousness and knowledge just sketched is highly speculative, and therefore not a very good basis for refuting the philosophy of knowledge. The following points, however, should be noted:

(a)   Even if speculative, the just sketched biological, evolutionary theory of the development of thought, reason, consciousness, meaning and knowledge in the world is at least itself coherent, intelligible. The decisive feature of this theory is that the ability to act successfully in the world, and to tackle rationally problems of action (at least to some extent) come into the world before explicit propositional knowledge – propositional knowledge instead being a development of the ability to act and to imagine actions. Thus it is certainly at least *intelligible* to assert that rational action is a precursor of, and not dependent upon, knowledge.

(b)   A major implication of the above theory is that animal learning is, in one important respect at least, from a methodological standpoint, more *rational* than our best official ideas about human learning – standard empiricism and the philosophy of knowledge. The philosophy of knowledge assumes that the prime task for inquiry is to acquire knowledge, which can then be applied to solving problems of living. Animal learning proceeds on the more enlightened principle that it is problems of action, of living, that are central and fundamental. However, we ought not to be surprised that we have much to learn from the methodology of animal learning. The ability to learn is clearly of great potential survival value: it is to be expected that the process of natural selection will select out only the most efficient, the most rational, learning methods.

(c)   The account given above of the development of knowledge is fully in accordance with our present scientific knowledge – in particular our knowledge of the physical universe and of evolution. In this respect it has a clear advantage over those accounts of the development of knowledge which take the philosophy of knowledge for granted, and thus require an essential break between animal and scientific learning. Thus Popper, presupposing a version of the philosophy of knowledge, is led to postulate an autonomous realm of propositions – his 'world 3' – which nevertheless somehow interacts with the brain; all this is difficult to accommodate within existing scientific knowledge of the physical universe and of life.

(d)   The above action-suspended theory of the development of thought, consciousness, meaning and knowledge in the world, though speculative, may well be true. If true, it provides a basis for arguing that it is not just that, as a matter of fact, the capacity to act and to imagine action have temporal priority over the capacity to develop propositional knowledge: rather these things are, as it were, rationally prior to the development of knowledge, necessary preconditions for knowledge to exist, so that the nature of knowledge cannot be understood at all unless understood as arising out of the capacity to act and to imagine action. Or, to put this more succinctly: if all our knowledge as a matter of fact has the character specified by the above theory, this provides a basis for maintaining that all knowledge must be conceived of as a development of the capacity to act successfully in the world.

But it must of course be admitted that the above theory, being empirical and speculative, does not provide a very good basis for arguing that all knowledge *must* be conceived as an aspect of, a development of, the capacity to act, to realize goals in the world. In seeking to establish this doctrine we must argue rather that all alternative views run into insuperable philosophical problems which this doctrine successfully resolves. In support of the doctrine, here are three further arguments.

(e) The ability to acquire knowledge requires the prior existence of the ability to act successfully in the world, and to imagine possible actions, since only in this case can there be any possibility of making observations and experiments needed in order to develop propositional knowledge (the presumption here being that knowledge-acquiring observation inevitably involves an element of action). On an individual level, it is only when we have discovered how to act successfully in the world as very young children, that there can be any possibility of acquiring propositional knowledge.

(f) Knowledge only exists if there exist conscious, self-aware, knowing persons; in order to be a conscious person it is essential to be able to act successfully in the world: hence successful action in the world is an essential precondition for the existence of knowledge. It might seem that a conscious but completely paralysed person constitutes a counter example to this argument. We may hold, however, that a completely paralysed person is only conscious insofar as it is legitimate to interpret his brain activity as being sufficiently analogous to brain activity which would occur if the person *could* act. Successful action in the world is even here hovering in the background, as it were, as a conceptual necessity for the person to be conscious.

(g) Propositional knowledge can only exist if people understand propositions: understanding propositions is itself, however, to be understood as a development of imagining actions (a development of imagining the environment for an action with the actor ignored); imagining possible actions in turn requires the prior existence of successful action. Thus an essential precondition for propositional knowledge to exist in the world is the existence of successful action.

The viewpoint developed here is fully in accordance with, and is backed up by, the arguments of chapters 4 and 5 concerning rationality, designed to establish that rational solving of problems of action presupposes a prior capacity to solve problems of action, to act successfully in the world, or that rational aim-pursuing presupposes a prior capacity to pursue aims.

Life, action, and the problems of life, of action, are not only historically and rationally (or conceptually) prior to knowledge and the problems of knowledge: in addition, I wish to argue, they are evaluatively prior. What really matters is what we do, what goes on in our lives. Knowledge is of importance insofar as it contributes to, and participates in, life.

I conclude that life, and the problems of life, are more fundamental than knowledge and the problems of knowledge, and that inquiry, rationally devoted to helping us achieve value in life must give priority to problems of life over problems of knowledge – a basic requirement which inquiry pursued in accordance with the philosophy of knowledge fails to satisfy.

Finally it may be objected that it is not at all clear what it means to assert that action and the problems of action are more fundamental than knowledge and problems of knowledge. Does this amount to a kind of idealism, somewhat like doctrines advocated by Hegel or Schopenhauer, according to which the world is somehow the product of our actions? The answer is no. The stars exist independently of our actions, our perception and knowledge of them. Indeed, in chapter 9 I argue that physical entities such as electrons and protons exist and have properties independent of our acts of observation and measurement. I advocate physical realism, even a doctrine that might be called conjectural essentialism. It is not the physical universe that is the product of human action: rather it is our knowledge of the physical universe that is such a product. All human knowledge, however ostensibly impersonal and formal, is the outcome of human action, presupposes successful action in the world, and needs to be pursued and communicated in such a way as to help enhance the value of life. Impersonally recorded knowledge, in libraries, exists in order to promote personal acts of knowledge, apprehension, exploration. Dissociated from life, it is just paper and ink.

Reply to objection 7
As to the objection that the philosophy of wisdom, unlike the philosophy of knowledge, cannot do justice to the value of inquiry pursued for its own sake, I claim the thing is all the other way round. Inquiry, pursued for its own sake, *is* people, individually and cooperatively, seeking to discover, understand and appreciate

that which is of value in existence, significant aspects of the world, as an end in itself. It can scarcely be distinguished from life itself. A life devoid of this dimension of searching for its own sake would be impoverished indeed. In chapter 4 I suggested that we can get an estimation of how highly we value our own personal inquiry pursued for its own sake by considering how highly we value our capacity to *see* for its own sake. To a very great extent, for all of us, the value of life is bound up in the value of exploring and discovering significant aspects of the world around us, for its own sake. It is just this that formal inquiry, pursued for its own sake, properly conducted, emerges out of and seeks to encourage. Science and scholarship, pursued for their own sake, are personal and passionate aspects of life, essential to the value of life. Even explorations into the furthest reaches of space, into the first few seconds of the cosmos, or into the fundamental laws of nature, into aspects of the world as far removed as possible from our customary human world and its concerns, nevertheless are of value insofar as they enrich human life.

All this can flourish, and can be understood, as long as inquiry is pursued in accordance with the philosophy of wisdom, since this viewpoint stresses the fundamentally personal and social character of inquiry, and stresses that inquiry has as its basic aim the realization of value in life – 'realization' including both discovery and creation. It cannot however flourish, or be adequately understood, as long as inquiry is pursued in accordance with the philosophy of knowledge, since this viewpoint dissociates the intellectual from the personal and social, and gives to inquiry the aim of acquiring impersonal knowledge of value-neutral fact. General adoption of the philosophy of knowledge even leads to a general blindness to the way in which modern science and scholarship, as a result of becoming institutionalized, professionalized, specialized, even industrialized, betray what is best, potentially, in inquiry pursued for its own sake, namely the shared, passionate quest of individuals into aspects of the mystery that surrounds us, and of which we are a part.

Our understanding and appreciation both of Nature and of ourselves (and of the interrelation between the two) are adversely affected by general adoption of standard empiricism and the philosophy of knowledge. In chapters 5 and 9 I attempt to demonstrate and illustrate how pursuing physical science in accordance with standard empiricism leads to a neglect of problems of understanding, the aim of science degenerating into the task merely to predict more and more phenomena more and more accurately. As a result, important problems of understanding that may not be all that hard to solve if put into the context of the endeavour to understand the nature of the physical universe – such as interpretative problems of quantum theory – remain neglected, misconstrued and unresolved.

Far more serious is the injustice that adoption of the philosophy of knowledge does to our understanding of each other, and of ourselves. As I have already pointed out, one can discern, in the history of thought, two apparently very different conceptions of understanding. On the one hand there is 'understanding' as this arises in the context of the physical sciences: here, we 'understand' some phenomenon to the extent that we can predict, and thus explain, the phenomenon by means of a comprehensive, unified and, ideally, true physical theory (the theory, at the very least, being predictively successful elsewhere). To explain, and thus to 'understand', means here to fit into a comprehensive pattern: and precise prediction is a necessary, though not a sufficient, condition for understanding to be achieved. On the other hand there is what might be called 'person-to-person' understanding, achieved when one person can accurately imagine himself to be another person, with that other person's feelings, desires, experiences, problems, beliefs, values. (Ideally, perhaps, this kind of person-to-person understanding is to be conceived of as a mutual affair, achieved by two people of each other.) It is this kind of person-to-person understanding that we find ourselves to be achieving of imaginary people, to a high degree, as we read or view great works of literature or drama, such as those of Shakespeare, Tolstoy or Chekhov. It is in this way that we seek to improve our understanding of our acquaintances, our friends, those we love. All good biographies, autobiographies, and some history, seek to improve our understanding of people, in this sense of 'understanding'. In contributing to the tradition of 'hermeneutics', 'verstehen' or 'empathetic understanding', thinkers as diverse as Vico, Herder, Ast, Wolf, Schleiermacher, Droysen, Dilthey, Croce and Collingwood have emphasized the importance of this kind of 'person-to-person' understanding, or have discussed or pursued aspects of it in their work. In recent times Laing (1965) and Sacks (1976) in particular have emphasized its importance for psychiatry and medicine.

However, as long as the philosophy of knowledge is accepted, 'person-to-person' understanding seems not only quite different from, but also intellectually inferior to, 'scientific' understanding. 'Scientific' understanding, it can be agreed, is (a) objective (b) impersonal (c) factual (d) rational (e) predictive (f) testable and (g) scientific, in that there is an objective, impersonal, factual theory, which predicts the phenomenon to be understood, and is independently testable, and so capable of being appraised scientifically and rationally. Any genuine example of 'person-to-person' understanding, based on one person imagining himself to be another person and thus seeing the other person's situation, experiences, problems, from his own point of view, is liable to lack all of the above features. Such an act of understanding is, it may well be argued, (a) subjective (b) personal (c) emotional and evaluative (and thus non-factual) (d) intuitive (and thus non-rational) (e) non-predictive (f) untestable (g) unscientific. In seeking to achieve 'person-to-person' understanding of another person in some particular context, I seek to put myself, imaginatively, into that person's shoes: I try to imagine myself to be him. In doing this inevitably I make all sorts of guesses about such things as the other person's feelings, desires, thoughts, problems, circumstances, beliefs and so on. But my primary aim is not to construct a *theory* at all, but rather to achieve an act of imaginative identification – to create within myself, in imagination, feelings, desires, aims, beliefs and so on, analogous in relevant respects to those experienced in reality by the other person. I may genuinely achieve this, and yet be unable to articulate a *theory* about the person in question, let alone a predictive, testable theory. Furthermore, I may not be able to predict the other person's actions or, if I am able to do so, I may well make a false prediction: and yet my 'person-to-person' understanding might be very good. I might, for example, realize that the other person faces a certain problem which he may seek to solve by doing one or other of three possible actions: all this might be entirely correct except that the other person hits upon a fourth action which had not occurred to me but which, once enacted, confirms my understanding of the person in that it constitutes an even better, or more typical, resolution of the problem confronting the person than the three possible solutions I had imagined.

Thus as long as something like the intellectual standards of the philosophy of knowledge are upheld, person-to-person understanding must be judged to be both quite different from, and vastly inferior to, 'scientific' understanding, from an intellectual standpoint. This ought to be conceded even by those who, like Winch, Bauman, Outhwaite, Giddens and Hesse, seek to defend the legitimacy or importance of some version of person-to-person understanding for the social sciences within the general framework of some version of the philosophy of knowledge. And, of course, any such apologetic defence of person-to-person understanding plays straight into the hands of those who argue that only 'scientific' understanding has any real intellectual merit.

When viewed from the standpoint of the philosophy of wisdom, however, all this is radically changed. The first point to note is that, in effect, according to the philosophy of wisdom, a central and fundamental task of inquiry is to promote the development of good person-to-person understanding between people in the world. For, according to the philosophy of wisdom, the cental and fundamental task of inquiry is to articulate problems of living experienced by people in their lives, and to propose and critically assess possible solutions – and to promote the doing of this in a cooperative way as an integral part of life. It is to help articulate and scrutinize life-goals, thus helping aim-oriented rationalistic ways of life to develop in the world. But it is just these things that we need to do in order to achieve person-to-person understanding of each other. If I am to enter imaginatively into another person's life, so that I imaginatively see and experience things from his point of view, the essential thing that I need to do is to recreate imaginatively the other person's life-problems and their possible solutions. I need to imagine that I have the other person's life-goals, shaped by his circumstances, past, temperament, skills, feelings, desires, values and beliefs, and that I am seeking to realize these goals in the kind of way in which he might (to put the matter in a rather more 'aim-oriented rationalistic' way). Thus the philosophy of wisdom in effect puts the development of person-to-person understanding at the heart of intellectual inquiry: inquiry pursued in accordance with the philosophy of wisdom is designed specifically to help us improve our person-to-person understanding of each other in life.

The second point is that the philosophy of wisdom, unlike the philosophy of knowledge, is able to do full justice to the supreme value of person-to-person understanding from both human and intellectual standpoints, here as elsewhere intellectual value reflecting and promoting that which is of human value. It is hardly too much to say – especially in the light of the arguments developed above – that almost everything of value in life depends in an essential way on people developing person-to-person understanding of each other. Certainly friendship and love depend on this. But more generally, we may argue, all cooperative action, and in particular all cooperatively *rational* action, depends upon people being able to develop this kind of understanding of each other. For if people are to be able to do things together cooperatively, taking joint responsibility for some shared, common enterprise, it is essential that those involved can 'understand' – can enter imaginatively into and identify with, at least to some extent – each other's problems, proposals, desires and goals. Without this, cooperative rationality cannot begin. The extent to which we can, or cannot, develop person-to-person understanding of each other is thus potentially of great social, political and moral importance. As the social world in which most of us find ourselves becomes increasingly vast, complex and diverse – so that increasingly we interact with others very different from ourselves – it becomes all the more important that person-to-person understanding can develop across such differences. Promotion of individual liberty, and thus of diversity of choices and ways of life at the individual level (itself a cooperative enterprise) is hardly likely to succeed or be sought if individual people cannot empathetically understand those different from themselves. The great danger is that in a vast, complex and diverse world people, instead of being enriched by diversity, will merely come to feel threatened and isolated by it, and will as a result hunger for some form of collectivism or nationalism (of the left or right) which banishes individual liberty and diversity. Freedom, justice, cooperative rationality, peace, active and effective compassion, friendship and love: all these depend quite essentially on the existence of person-to-person understanding between people. And, so we may argue, person-to-person understanding is essential to the realization of value in life in even more basic ways than this, in that it is essential to all communication between people, essential to our development as persons, essential to the development of self-consciousness. Quite generally, we are able to realize things of value in our life because of imaginative identification with the value-realizing endeavours of others.

As long as the intellectual merit of understanding is assessed in terms of criteria appropriate to the philosophy of knowledge, in terms of such things as the factual content and predictive power of impersonal, objectively formulated theories, person-to-person understanding cannot amount to very much from an intellectual standpoint, whatever its human value may be. Acceptance of the philosophy of wisdom, however, leads us to assess the intellectual merit of 'understanding' in a different way, in terms of how important and central it is in helping us to realize what is of value in life. Assessed in this way, person-to-person understanding emerges as being of profound intellectual value and merit. In connection with the above seven apparent intellectual defects and disadvantages of person-to-person understanding the philosophy of wisdom enables us to say the following. (a) Person-to-person understanding can be, and ought to be, wholly objective in that it does full justice to the actual (objectively existing) feelings, desires, beliefs, aims and values of the person to be understood. In addition it can be, and ought to be, objective in that it includes knowledge of the person's actual circumstances, actions and aims, and knowledge of what is genuinely of value to him, actually and potentially, in his circumstances, as opposed merely to what the person himself believes about all this. One does not understand another person merely by sharing that other person's illusions and delusions. (b) Person-to-person understanding is certainly a *personal* (and inter-personal) kind of understanding, but none the worse for that. For certain purposes some personal aspects of thought may be neglected: it ought always to be remembered, however, that all thought is in the end personal and inter-personal, pursued for personal and social ends. (c) Person-to-person understanding certainly does in general include the imaginative sharing of feelings and desires: but far from being non-factual, such understanding, to be any good, must involve knowledge and understanding of facts having to do with the circumstances of the person to be understood; and feelings and desires imaginatively experienced must in fact be similar to those of the person being understood. Our understanding of others *ought* to involve imaginative experiencing of their feelings and desires, and the capacity – or at least the concern – to see what is of value, potentially and actually, in the circumstances of their lives. Our mutual development of this kind of understanding of each other enriches us all: in its absence we are all impoverished. (d) Being able to achieve good understanding of others is to some extent a skill which, like other skills – such as those involved in speaking a language, for example, or in doing scientific research – can be acquired as a result of desire and practice. In successfully using an acquired skill we may act spontaneously, instinctively or intuitively; that is, without reflection, and not quite understanding why we act as we do: and yet our performance may be entirely rational, learnable and in accordance with sound general (but implicit) principles or methods – in this case the principles of the philosophy of wisdom. (e) Person-to-person understanding is not of intellectual value insofar as it involves seeing apparently disparate phenomena as aspects of a precise, comprehensive pattern, and is not of practical value because it enables us to predict, and therefore control:

rather, it is of intellectual and practical value insofar as it involves seeing ourselves in others (and others in ourselves), thus making possible cooperative rational action, communication, sympathy, friendship and love. Person-to-person understanding is not encapsulated in impersonally formulated predictive and explanatory theories: but this does not in itself cast doubt on the intellectually sound, important and fundamental character of this kind of understanding since, according to the philosophy of wisdom, problems of living and their possible solutions (possible actions) are at the heart of rational inquiry rather than problems of knowledge and their possible solutions (theories). (f) In seeking to improve our person-to-person understanding of others it is essential to proceed in a conjectural and critical way, 'testing' our understanding by means of communication, listening to the testimony of others, considering possible implications for ourselves, for our own actions, of having certain feelings and desires, checking the network of factual assumptions implicit in such understanding. (g) It is absurd to condemn person-to-person understanding as non-rational and unscientific since such understanding is essential to all cooperatively rational pursuits, and in particular essential to the cooperatively rational pursuit of science. In order for 'scientific' understanding of natural phenomena to be developed, it is essential that scientists can develop person-to-person understanding of each other. Impersonal, public, scientific knowledge, resting on a multitude of agreements among scientists about meanings, methods and results, is the outcome of a long history of individual scientists seeking to 'understand', in a person-to-person sense, each other's ideas, problems, projects, objectives, proposals. It is in this way that 'scientific' understanding of natural phenomena is based on person-to-person understanding between scientists. That which one scientist seeks to achieve in attempting to understand the theoretical or experimental work of a colleague is not essentially different from what a historian of science seeks to achieve in attempting to understand the work of Galileo, Newton or Darwin; and this in turn is not essentially different from what anyone seeks to achieve in acquiring person-to-person understanding of the work, life or actions of any other person, living or dead. Imaginatively reorganizing the way I see the world so that it more nearly resembles the way you see the world (which I must do if I am to be able to have person-to-person understanding of you) is not essentially different from imaginatively reorganizing the way I see the world so that it more nearly resembles the way Einstein saw the world, or intended us to see the world, in propounding the special or general theory of relativity. In both cases I am concerned with possible imagined visions of the world (more or less comprehensive, precisely formulated, testable, empirically successful, and so on). In both cases the emphasis of my interest may be *personal* (I wish to improve my understanding of you, or of Einstein), or *impersonal* (I wish to improve my understanding of the world). However, these two sorts of interest ought not to be, and strictly cannot be, severed from one another. If in pursuing theoretical physics we lose sight of the fact that our ostensibly 'impersonal' understanding of natural phenomena is the outcome of persons sharing, criticizing and developing each others' personal imaginings (and thus an outcome of person-to-person understanding between physicists), we are likely, as we have already seen, to betray the intellectual heart of the whole enterprise of natural philosophy: if in enhancing person-to-person understanding of each other we abandon our best cooperative efforts at improving our knowledge and understanding of the world, and give equal validity to all world views, we descend into mere relativism and subjectivism, and abandon the means to distinguish between sanity and insanity, justice and injustice, democracy and totalitarianism.

The two kinds of understanding dovetail together, being interdependent. Only the philosophy of wisdom can do justice to both kinds of understanding, and to their interdependence, in a unifying way, both being essential to wisdom.

Reply to objection 8
Utopian social engineering, as described and criticized by Popper, 'aims at remodelling the "whole of society" in accordance with a definite plan or blueprint; it aims at "seizing the key positions" and at "extending the power of the State . . . until the State becomes nearly identical with society", and it aims, furthermore, at controlling from these "key positions" the historical forces that mould the future of the developing society' (1961, p. 67). Popper formulates the basic argument appealed to by Utopians to support this plan of action as follows:

> Any rational action must have a certain aim. It is rational in the same degree as it pursues its aim consciously and consistently, and as it determines its means according to this end. To choose the end is therefore the first thing we have to do if we wish to act rationally; and we must be careful to determine our real or ultimate ends, from which we must

distinguish clearly those intermediate or partial ends which actually are only means, or steps on the way, to the ultimate end. If we neglect this distinction, then we must also neglect to ask whether these partial ends are likely to promote the ultimate end, and accordingly, we must fail to act rationally. These principles, if applied to the realm of political activity, demand that we must determine our ultimate political aim, or the Ideal State, before taking any practical action. Only when this ultimate aim is determined, in rough outline at least, only when we are in possession of something like a blueprint of the society at which we aim, only then can we begin to consider the best ways and means for its realization, and to draw up a plan for practical action. These are the necessary preliminaries of any practical political move that can be called rational, and especially of social engineering. (1969, vol. 1, pp. 157–8)

There are, perhaps, some Utopian aspects to what is being advocated in this book. I do seek to help remodel the 'whole of society'. I do, after all, propose that we should seek to develop a cooperatively rational world society. As a means to this end, I propose a sweeping, holistic change in the overall aims and methods of institutionalized inquiry and education, from knowledge to wisdom. Organized, academic inquiry, I am arguing, needs to take as its basic intellectual aim to help us develop a cooperatively rational, world society. Furthermore, I have argued that the holistic, Utopian institutional change that I am advocating in connection with academic inquiry can be taken as a model, for all of social inquiry, as to how we should seek to transform all other institutions in the world.

Despite this, it is also clear that what is advocated in this book is quite different from Utopian social engineering, as characterized by Popper. Cooperative, rational, social problem-solving, as characterized here, involves action that is very different from 'seizing the key positions', or 'extending the power of the State until the State become nearly identical with society'. Aim-oriented rationalism involves articulating aims certainly: but it does not assert that 'to choose the end is the first thing we have to do if we wish to act rationally'. Quite the contrary, aim-oriented rationalism asserts that if we wish to act rationally we must seek to improve our aims and methods as an integral part of *what we are already doing*. We act rationally when we add to the aim-pursuing we are already engaged in the activity of imagining and criticizing possible and actual aims and methods, in an endeavour to discover how our actual aims and methods may be improved, little by little, as we proceed.

Popper contrasts Utopian engineering, which he rejects as irrational and disastrous, with piecemeal social engineering, which he advocates as rational and humanitarian. In *The Poverty of Historicism* Popper characterizes piecemeal engineering as follows:

> Even though he (the piecemeal engineer) may perhaps cherish some ideals which concern society 'as a whole' – its general welfare perhaps – he does not believe in the method of re-designing it as a whole. Whatever his ends, he tries to achieve them by small adjustments and re-adjustments which can be continually improved upon. His ends may be of diverse kinds, for example, the accumulation of wealth or of power by certain individuals, or by certain groups; or the distribution of wealth and power; or the protection of certain 'rights' of individuals or groups, etc. Thus public or political (piecemeal) engineering may have the most diverse tendencies, totalitarian as well as liberal . . . The piecemeal engineer knows, like Socrates, how little he knows. He knows that we can learn from our mistakes. Accordingly, he will make his way, step by step, carefully comparing the results expected with the results achieved, and always on the look-out for the unavoidable unwanted consequences of any reform; and he will avoid undertaking reforms of a complexity and scope which make it impossible for him to disentangle causes and effects, and to know what he is really doing. (1961, pp. 66–7)

And Popper adds 'Once we realize . . . that we cannot make heaven on earth but can only improve matters a *little*, we also realize that we can only improve them *little by little*' (1961, p. 75).

In *The Open Society and its Enemies*, the aim of piecemeal engineering is characterized somewhat differently, in that Popper there asserts that the piecemeal engineer

> will be aware that perfection, if at all attainable, is far distant, and that every generation of men, and therefore also the living, have a claim; perhaps not so much a claim to be made happy, for there are no institutional means of making a man happy, but a claim not to be made unhappy, where it can be avoided. They have a claim to be given all possible help, if they suffer. The piecemeal engineer will, accordingly, adopt the method of searching for, and fighting against, the

greatest and most urgent evils of society, rather than searching for, and fighting for, its greatest ultimate good. (1969, vol. 1, p. 158)

Granted that a choice must be made, then clearly cooperative, rational, social problem-solving, as characterized in this book, is very much more like piecemeal than Utopian social engineering, especially as piecemeal engineering is characterized in *The Open Society and its Enemies*. But there are still important differences between the two (even if the moral neutrality of piecemeal engineering as characterized in *The Poverty of Historicism* is ignored). The method of piecemeal engineering does not appear to be, fundamentally, a method of *cooperative*, social problem-solving – that is, a method whereby many people cooperatively take responsibility for and guide social problem-solving together. The method of social engineering fails to emphasize the vital need to try out many possible individual and social actions in the individual and social imagination, so that unforeseen undesirable consequences of actions may be discovered in the *imagination*, and not in reality. In particular, the method of social engineering fails to emphasize the vital need for individuals cooperatively to articulate and criticize proposals for perhaps quite radical cooperative social change, so that many individuals can discover for themselves what the consequences would be, for their own individual lives and problems, of taking part in the cooperative *enactment* of such proposals; and so that personal problem-solving can acquire a cooperative social context and perspective. For a group of people, or a society, to carry out cooperatively some plan of action, it is not necessary for any individual, or small group of individuals to be able to understand and anticipate all the detailed problems for individuals that carrying out the plan would create: all that is necessary is that individuals can anticipate and tell others when such problems become insoluble or insufferable, so that in these cases the overall plan can be appropriately and cooperatively modified. It is just such potential rich resources of cooperative rational problem-solving as this that Popper's piecemeal engineering ignores. In so harshly criticizing Utopian engineering Popper leaves no room for the imaginative proposing and criticizing of *proposals* for radical but cooperative social change: it is perhaps this, more than anything else, which differentiates Popper's piecemeal social engineering from cooperative, rational, social problem-solving as described and advocated in this book.

Cooperative, rational, social problem-solving is, I suggest, in the light of these considerations, a *third* method for solving social problems, distinct from and superior to both Utopian and piecemeal social engineering, and not envisaged by Popper.

Two important general points deserve to be made about radical, cooperative social change – the kind of change Popper fails to envisage, or holds we should not attempt to make.

In the first place it is important to understand why it is so especially difficult for modern societies – for the modern world – to bring about generally desired and desirable changes of this kind. It is important to understand, that is, the nature of the problems that arise in an especially acute form for us in the modern world in seeking to implement radical cooperative social change. Engaging in cooperative social action is, of course, confronted by many problems today that are essentially the same as the problems that have always confronted such social action throughout history and pre-history: how to combat the powerful, the criminal and the deceiving who seek to oppose or subvert cooperative action; how to reach general agreement about what policies to adopt, what changes to attempt to make. Even a tribal society, a relatively small group of people, may be confronted by problems of this nature. The problems that confront us in the modern world, in an especially severe form, in addition to the traditional problems of cooperative action, are, I suggest, essentially of a *logistic* character. Modern societies are very much bigger, more complex, specialized and diversified than societies of the past. In pre-historical times, people lived in small cohesive hunting and gathering tribes. No logistic problem arises in calling a tribal meeting to discuss and to decide how to solve, in a cooperative manner, some problem that confronts the tribe. Furthermore, in comparison with the modern world, many features of tribal life will tend to promote in individuals attitudes conducive to assuming shared, cooperative responsibility for the welfare of the tribe. All the members of the tribe are known to each other personally. Relationships of mutual interdependence are experienced daily, on a personal basis, in hunting, gathering food, and so on. Obligations and responsibilities towards fellow members of the tribe can be experienced in a personal, emotional way, in terms of known individuals in much the same way as we can experience responsibilities towards members of our family today. All members of the tribe have a common outlook on things, a common cosmology, religion, system of values: thus barriers to intimacy, to mutual understanding, do not arise as a result of differences of outlook and values. Because of the relative smallness of the tribe, each individual makes a personal impact on the life of the tribe as a whole, and can be well aware of this impact. The tribe, as it were, acknowledges the existence, value and potency of the individual, and is clearly affected by the actions of the individual. It will be normal, and not abnormal, for the individual to suppose he or she has sufficient personal importance in the life of the tribe to have a role in taking cooperative responsibility for the life of the tribe.

Time passes; agriculture is invented; trade grows and tribes coalesce; work becomes more and more diversified and specialized; modern methods of travel and communication are invented and developed. As history unfolds, the tribe becomes the nation, and the nation the world.

As a result of these rapid changes (incredibly rapid if. put into the context of the three and a half million years or so mankind has been in existence on earth), the problems of acting cooperatively on a world-wide basis have become immense. From the standpoint of the world-wide community of humanity, each individual is surrounded by millions upon millions of complete strangers, most of whom speak languages, live lives, uphold beliefs and values that are more or less incomprehensible. Almost all individuals are utterly powerless to have any sort of impact on national life, let alone on world society. Opportunities for discovering how to take cooperative responsibility for the modern world will for most simply not arise. Almost everything tending to promote cooperative action in tribal life has, in the modern world, disappeared. Indeed some of the developments we most prize, such as the development of specialization, and diversity of ways of life, the freedom of individuals to adopt beliefs, values, modes of being different from others, serve to increase the severity of the problem.

There are, in short, utterly obvious, wholly unmysterious, essentially *logistic* reasons as to why cooperative action has become so severely problematic in the modern world.[3]

As I indicated in chapter 4, a fundamental task of organized inquiry, according to the philosophy of wisdom, is to help provide solutions to the logistic problems of engaging in cooperative action in the modern world. According to the philosophy of wisdom, we need to develop academic inquiry as a sort of institutional surrogate for open, cooperative tribal meetings of humanity.

---

[3]Popper might agree with some of this. His failure, however, to envisage the possibility of the philosophy of wisdom, making cooperative action possible even in an 'open' society, leads him to hold that 'the strains of civilization' must be endured rather than resolved.

The second general point to note about radical, cooperative, social change is that circumstances can develop which make it all but imperative to make such changes, irrespective of the size of the community involved. A small tribal society may need to change quite drastically a way of life that has been established for generations – because of some change in the circumstances of life such as a change of climate, the advent of hostile neighbours or a new disease, or dwindling traditional food resources. Equally, a vast, complex, world society may need to change, relatively suddenly and drastically, complex, diverse ways of life that have been established for generations – because of a relatively sudden, world-wide change in the circumstances of life.

We are confronted today by just such problems, on a world-wide basis. The population explosion, the world-wide endeavour to industrialize and achieve rapid economic growth, the resulting rapid depletion of finite natural resources and rapid destruction of plant and animal life, and the development of the nuclear bomb, taken together constitute a sudden, dramatic change in the world-wide circumstances of life for humanity.

In the light of these points, it is clearly a matter of extreme urgency that academic inquiry, on a world-wide basis, begins to give intellectual priority to the tasks of (1) improving the articulation of these problems and (2) putting forward and criticizing proposals as to how these problems might be cooperatively resolved – thus promoting cooperative rational, social problem-solving in the world by putting the philosophy of wisdom into practice.

But this violates the principles of Popper's piecemeal engineering. We have here an indication of the drastic inadequacy of what Popper prescribes.

What can have induced Popper to overlook the possibility of attempting to develop cooperative, rational, social problem-solving? The explanation is simple enough and will by now be very familiar. Popper upholds the philosophy of knowledge. For him, social inquiry is to be developed as social *science*, on analogy with natural science. Popper does, it is true, emphasize the importance of attempting to help solve practical problems in developing social science: but, as he goes on to point out, this does not differentiate social science from physical or biological science, since here too a concern for practical problems is important, not only because solving practical problems may be important in itself, but also because concern for practical problems may stimulate progress in theoretical scientific knowledge in that it is 'invaluable for scientific speculation, both as a spur and as a bridle'. Thus 'Pasteur's reform of the biological sciences was carried out under the stimulus of highly practical problems, which were in part industrial and agricultural' (1961, p. 56). Popper goes on to argue that social science needs to be developed as *piecemeal social technology*, an adjunct to piecemeal social engineering, as a collection of scientific social laws which specify what *'cannot be achieved'* (1961, p. 61). Popper gives a list of candidates for social laws of this type, and argues that the laws and theories of natural science are of precisely the same form, in that 'every natural law can be expressed by asserting that *such and such a thing cannot happen* . . . For example, the law of conservation of energy can be expressed by: "you cannot build a perpetual motion machine" ' (1961, p. 61). And Popper remarks that 'the significance of our analysis lies in the fact that it draws attention to a really fundamental similarity between the natural and the social sciences' (1961, pp. 61–2).

Thus for Popper, the task of social science is to improve our knowledge of testable social laws prohibiting certain sorts of social actions, the hope being, presumably, that it will be possible, eventually, to develop unifying testable social theories, from which a wide range of social laws will be derivable just as a wide range of natural laws are derivable from the unifying, testable theories of physical science. From this standpoint, social science must be judged to be in an incredibly primitive state. This is stressed by Popper, as when he remarks 'My point about the technological approach might perhaps be made by saying that sociology (and perhaps even the social sciences in general) should look, not indeed for "its Newton or its Darwin", but rather for its Galileo, or its Pasteur' (1961, pp. 59–60).

It is especially the primitive state of social science, of social technology, so necessary, from Popper's standpoint, for piecemeal social engineering, that leads Popper to condemn the attempt to make radical, far-reaching social changes, and to condemn even, by implication, the enterprise of putting forward and criticizing proposals for such social changes. Thus he says: 'At present, the sociological knowledge necessary for large-scale engineering is simply non-existent' (1969, vol. 1, p. 162).

Popper's line of argument has the effect of prohibiting the one social change that is now so urgently needed if humanity is to discover, little by little, how to tackle its common problems in more cooperative and humane ways – namely a change in academic inquiry, and above all in social inquiry, from knowledge to wisdom. Popper fails to make the two general points about radical, cooperative, social change stressed above, namely that such radical social change may well be urgently needed, and that the peculiar difficulties that confront us in seeking to bring about such social change are essentially logistic in character (as this was interpreted above). Failing to construe the problems in this way, he also fails to put forward the solution advocated here – namely that organized inquiry should be developed precisely in order to promote cooperative, rational, social problem-solving in the world. His allegiance to the philosophy of knowledge blinds him to this even as a possibility. His view that social problem-solving is at present severely hampered by the primitive state of social science, and his resulting advocacy of the urgent need to develop social inquiry as the pursuit of social knowledge, as social science (on analogy with natural science) actually serves to prohibit the one thing that most urgently needs to be done if we are to solve our common problems of living in a more humane fashion: namely develop academic inquiry in accordance with the edicts of the philosophy of wisdom. His arguments against Utopianism in our present state of sociological ignorance, has the effect of prohibiting just the kind of thinking we need to engage in if we are to solve our problems more cooperatively, rationally and humanely.

According to the philosophy of wisdom, it is a fundamental intellectual obligation of every teacher, every social inquirer, every scientist and scholar, in his or her professional work, to put forward and criticize proposals for cooperative action intended to promote the realization of what is of value in life and to encourage others to do this. These proposals may be limited and specific; or they may be unrestricted as one can imagine, in that they may well amount to proposals as to how humanity as a whole is to achieve 'heaven on earth'. The vital point is to promote in society the *habit* of putting forward and criticizing proposals for action intended to help achieve what is of value. Only a society which has this habit can engage in cooperative, rational, social problem-solving and action. A reader of Popper's diatribes against Utopianism, however, is likely to conclude that to put forward for consideration and criticism sweeping, holistic proposals for social action – as I have done in this book – is to commit the deplorable intellectual, political and moral sin of advocating Utopian social engineering. He or she will thus refrain from putting forward such proposals, and will discourage others from doing so. In this way, Popper's critique of Utopianism has the effect of sabotaging the enterprise of developing a more cooperatively rational society, even if this is not exactly the effect Popper intended. It should be noted that Popper here joins company with Marx: for Marx's diatribes against Utopian socialism must tend to have exactly the same effect.

From the standpoint of the philosophy of wisdom, Utopians are objectionable in so far as they put forward Utopian proposals that exclude cooperative living and problem-solving; dogmatically uphold such proposals; attempt to implement such proposals in uncooperative ways. Many Utopians commit some, if not all, of these sins. But this must not lead us to the conclusion that all Utopians must commit some, if not all of these sins. For then we prohibit cooperative reason.

Reply to objection 9
The view that individual freedom and cooperativeness are inherently at odds with one another is a mistake. Quite to the contrary, freedom of individuals and cooperation between individuals are mutually interdependent. Cooperation between individuals depends on individual freedom simply because without individual freedom there can be no cooperation (as 'cooperation' is understood in this book). Individuals only cooperate with one another if they do so freely, as a result of their own desire and decision to do so: to the extent that one group is tricked, bullied or brainwashed into acting 'cooperatively' by another group, to that extent cooperation does not take place. Cooperation is essentially free and mutual: for A and B to cooperate it is essential that A cooperates freely with B and B with A. Then again, freedom of individuals depends on cooperation between individuals, for two general reasons. First, in our crowded, complex and interdependent world, without cooperation, freedom of individuals will collide. One person's exercise of freedom will demand another person's loss of freedom. One group's exercise of freedom will demand another group's loss of freedom. Second, our individual freedom depends on our capacity to realize what is of value in life. Much that is desirable and of value in life – such as friendship and love – is itself cooperativeness going on between people of various kinds. Thus freedom requires cooperation. In short, cooperation is implicated in almost everything of value (and thus necessary to freedom) both as a *means* to the realization of things of value, and as something of supreme value in itself, an *end* in itself.

It follows that if our concern is to help enhance individual liberty (as a part of our concern to help achieve a more civilized world), it is essential that we strive to help develop a more cooperative world. A basic defect of traditional liberalism is that it has failed to emphasize the fundamental importance of developing a more cooperative world and has even, in some ways, actually opposed this. Traditional liberalism sees the problem of individual freedom as, fundamentally, the problem of defending the weak individual against the strong group, government or nation. From this standpoint, either we do what the group or government wants us to do, thus cooperating and abandoning our individual freedom, or we stand up against cooperation, thus preserving freedom. The programme of developing a more cooperative world is thus understood to be a programme for the annihilation of individual freedom, a programme for some kind of collectivist enslavement.

It is correct to see the problem of protecting the weak individual against the strong group or state as a basic problem of individual liberty. It is correct to oppose 'cooperation' in the sense of enslavement to group will. The *mistake* is to adopt, and restrict one's attention to, anti-liberal, collectivist conceptions of 'cooperation' alone. The moment cooperation is understood as presupposing individual freedom, it becomes clear that developing a more cooperative world, in this liberal sense, is not only compatible with, but is actually essential for, the development of individual freedom. The freedom to cooperate, one might almost say, is *the* freedom. In a world in which mutual cooperation and good will can be taken for granted, individual freedom is vastly enhanced for all of us, in all sorts of ways. Protecting weak individuals against strong groups or governments must itself be a cooperative act. Without cooperativism, liberalism collapses, intellectually and in practice.

CHAPTER NINE

Refutation of Minimal Standard Empiricism: From Science to Natural Philosophy

If organized inquiry is to help us realize what is of value in life in a good, rational way, then its overall character must be, in many respects, much closer to the philosophy of wisdom than to the philosophy of knowledge. This much at least, I trust, has now been established.

However, even granted this, a last ditch attempt at a defence of a highly modest version of the philosophy of knowledge may still be made. 'It is a grotesque mistake' (so it may be argued in support of this defence) 'to interpret the philosophy of knowledge as applying to the *whole* of rational inquiry. Such an interpretation was never intended in the first place. Properly understood, standard empiricism and the philosophy of knowledge are to be interpreted as applying, much more modestly, only to an *aspect* of that fragment of rational inquiry that is devoted to the acquisition of knowledge. Much of rational inquiry may well have other goals and may be concerned with problems of action: insofar as the acquisition of knowledge is not here at issue, such branches of inquiry lie beyond the intended scope of the philosophy of knowledge.

'It is true' (so the argument may be continued) 'that some modifications must be made to the somewhat primitive doctrine expounded in chapter 2, in the light of some of the arguments of subsequent chapters. It is, for example, wrong to exclude discussion of aims and priorities of research from the intellectual domain of science – from journals, lectures, textbooks and so on. There may well be fallible methods of discovery in science, which cannot be properly exploited if possible aims for research do not receive explicit scientific discussion. Again, the important role that *values* play in science needs to be recognized: of course science endeavours to discover what is significant or useful, and not only what is true, however irredeemably trivial.

'Despite such concessions to the philosophy of wisdom, the basic tenets of standard empiricism and the philosophy of knowledge, properly interpreted, remain entirely valid. In science questions about the *value* of a potential contribution to science must be sharply dissociated from questions about *truth*, about verification and falsification. As far as the latter questions are concerned, ultimately only empirical considerations must be allowed to determine what potential contributions are to be accepted and rejected as scientific knowledge. And more generally, only matters having to do with truth, fact, knowledge, must be allowed to influence what is accepted and rejected as constituting knowledge. All personal and social aims, problems, feelings, desires, experiences and values must be ruthlessly ignored when it comes to deciding what is to be accepted and rejected as constituting knowledge. To this extent, the basic tenets of standard empiricism and the philosophy of knowledge remain wholly valid.'

In this chapter I set out to establish that even this highly modest version of the philosophy of knowledge, which makes so many concessions to the philosophy of wisdom, is unacceptable, and must be rejected. Empirical considerations alone cannot decide what theories are to be accepted and rejected in science: metaphysical considerations concerning the comprehensibility of the universe must be taken into account in addition to empirical considerations. Furthermore, ignoring all personal experiences, feelings and desires when it comes to the assessment of claims to knowledge severely restricts what we can acquire knowledge about. If knowledge is not to be severely restricted, we must attend to our personal feelings and experiences in assessing many claims to knowledge. I discuss this last point first.

A major objection to the modest version of the philosophy of knowledge just outlined is that it necessarily excludes from inquiry all knowledge of just that aspect of the real world which gives life its meaning and value. As I shall argue in chapter 10, what is of value in existence is to be associated with what may be called the 'experiential' aspect of reality. The experiential has to do with all that we personally experience – with what we see, hear, smell, touch, feel, with what we become aware of through our own sensory and emotional responses to things. It includes the sensory and aesthetic qualities of things and works of art, and the personal and moral qualities of people and their actions. In chapter 10 I expound a view which I call *experiential realism*, according to which these experiential qualities of things and people do really exist in the objective world of fact, even though we can only become aware of them via our own personal experiences, our own personal sensory and emotional responses to things. Thus a rose really is red, as perceived by us, even though in order to become aware of this perceptual quality it is necessary oneself to experience the visual sensation of redness, it being impossible for a person blind from birth to know what sort of quality experiential redness is. Similarly, personal qualities of people we inadequately characterize by means of such terms as shyness, cynicism, courage, generosity, deviousness and so on, are real qualities of people and their deeds, even though we only become aware of these qualities via relevant personal experience.

In contrast to the experiential, there is the non-experiential dimension of reality – that aspect of things which one can know and understand without it being necessary oneself to have any special kind of experience. Thus, in order to understand what physical properties such as mass or elasticity are, it is certainly necessary to have had some experiences, simply in order to be conscious and capable of knowing and understanding anything at all. There is, however no particular sort of experience that it is necessary to have had. A congenitally blind or emotionless person may be able to understand all of physics just as well as a sighted or feeling person can.

All knowledge of this experiential realm – so vital from the standpoint of realizing what is of value in life, from the standpoint of acquiring wisdom – is however excluded *a priori* from inquiry pursued in accordance with the modest version of standard empiricism and the philosophy of knowledge as outlined above (and from the immodest versions expounded in chapter 2). In order to know experiential facts about things or people it is essential to attend to one's own personal sensations and emotions. According to standard empiricism and the philosophy of knowledge, however, personal sensations and emotions have no role to play whatsoever in assessing claims to knowledge. Only impersonal, de-sensorized, de-emotionalized observation and experimentation are relevant to the assessment of claims to knowledge. Thus inquiry pursued in accordance with the philosophy of knowledge (in its modest or immodest version) is necessarily restricted to improving our knowledge of non-experiential fact. Knowledge of experiences, feelings and sensations of people may of course be acquired in psychology, anthropology, sociology or history: this will, however, be knowledge of *non-experiential* aspects of experiences, sensations and feelings.

When the world is viewed through the spectacles of the philosophy of knowledge, all that is of value, all that makes life worth living, mysteriously vanishes, and leaves not a wrack behind. The beauty of a summer's day, the joy of a child's laughter, the miracle of being alive and conscious: all such things of value fade into mere non-experiential fact. The miracle of value realized, and the tragedy of value unrealized – whether through death, poverty, suffering, disease, enslavement or other misfortune – dissolve into non-miraculous, non-tragic, non-experiential fact. That there is anything of value in existence – anything miraculous or tragic – can only be, from the standpoint of even the modest version of the philosophy of knowledge, some kind of subjective illusion, an absurd hallucinatory reaction having nothing to do with objective reality, with objective fact.

When we are young (and if we are fortunate), the world is rich with value, charged with sensory and emotional significance, vivid with colour, taste, feel, smell, mystery, joy, terror, pleasure and pain. We are raw and open to the experiential dimension of reality.

According to the philosophy of wisdom, the basic task of rational inquiry and education is to help us strengthen and deepen this precious childish rawness and openness to the experiential dimension of reality. The task is to help us develop our childish capacity to realize what is of value so that it gradually becomes more sensitive and realistic, more knowledgeable and understanding, more creative, cooperative and responsible, more loving. In so far as this involves improving knowledge and understanding, it is above all knowledge and understanding of what is of value in the experiential realm that is of importance.

All this stands in sharp contrast with what is achieved by rational inquiry and education pursued in accordance with the philosophy of knowledge (even in its modest version). Knowledge and understanding acquired in accordance with precepts of the philosophy of knowledge do not strengthen and deepen our childish openness to the experiential: quite to the contrary, if anything, they quietly annihilate it. For, such knowledge and understanding carries with it the implicit and powerful message that our personal sensory and emotional reactions to things are irrelevant when it comes to a determination of objective fact. Becoming knowledgeable and educated actually involves becoming blind to all that is of value in the experiential realm – in that this realm is excluded from the world of objective fact and knowledge. Awareness of what is of value in the experiential realm is relegated to the merely subjective and personal, the illusory and non-rational, thus being decisively split off from the realms of objective knowledge and fact.

From the standpoint of the philosophy of wisdom this represents, not education, but a kind of intellectual corruption, the progressive inculcation of extreme intellectual blindness and schizophrenia. Is it to be wondered at if such value-blind knowledge, increasingly influential in the twentieth century, should on occasions be associated with such human horrors as the nuclear devastation of Hiroshima and Nagasaki, or the Vietnam war? Is it to be wondered at that those who are subjected to such value-blind education should end up puzzled and uncomprehending as to what really is of value in existence, and how it is to be cooperatively achieved?

Proponents of the philosophy of knowledge are, of course, absolutely correct to hold that our personal sensations, feelings and desires can often mislead us about the real objective nature of things. The disastrous mistake is to hold that our personal sensations, feelings and desires *always* mislead us about the real, objective nature of things. If we are to improve our knowledge and understanding of what is of value in existence (let alone improve our capacity to help realize what is of value), it is vital that we educate our sensory, emotional and motivational reactions to things, so that gradually these reactions may come to represent more faithfully to us objective experiential facts as opposed to emotional (or value) illusions and hallucinations. It is just this rational education of feeling and desire which becomes incomprehensible once one adopts the philosophy of knowledge view that all personal emotional and motivational responses to things are subjective and illusory.

Our personal capacity to realize what is of value in our life depends vitally on our own instinctive personal, emotional, and motivational responses to things becoming educated to reveal to us what is genuinely of value (to us) in existence. Our common capacity to develop a more civilized world depends vitally on our common emotional and motivational responses to things becoming educated to reveal to us what is genuinely of value in existence. The philosophy of wisdom encourages, and the philosophy of knowledge discourages, this vital kind of emotional and motivational learning.

In reply to these points, proponents of the philosophy of

knowledge may deny that the experiential realm really does exist: the truth of experiential realism may, in other words, be denied. (Arguments in support of experiential realism are developed below in chapter 10.) Alternatively, it may be conceded that the precepts of the philosophy of knowledge apply only to the acquisition of a highly restricted kind of knowledge, namely knowledge of non-experiential fact. I turn now to a refutation of this excessively modest version of the philosophy of knowledge in a domain as remote as possible from the experiential, where it ought to meet with its greatest success – namely the domain of theoretical physics.

According to the excessively modest, last remaining fragment of, standard empiricism and the philosophy of knowledge, now under consideration, theoretical physics at least obeys what Popper has called '*the principle of empiricism*, which asserts that in science, only observation and experiment may decide upon the *acceptance or rejection* of scientific statements, including laws and theories' (1963, p. 54). The basic aim of physics, in the context of justification, is to increase knowledge about the physical universe, no permanent presupposition being made about the nature of the universe. (If physics were to make any kind of permanent presupposition about the nature of the universe, then all those theories in conflict with this presupposition would be rejected out of hand, whatever their empirical success might be: this would involve violating the principle of empiricism, in that evidence *alone* would not decide what theories are to be accepted or rejected.) Optimistically, we may hope that knowledge of truth is gained at the *theoretical* level – successive theories drawing closer and closer to the truth; pessimistically, we may only require, for scientific progress, that knowledge of truth is gained at the *empirical* level – successive theories merely predicting more and more phenomena more and more accurately.

It should perhaps be noted that most contemporary scientists and historians and philosophers of science, including Carnap, Hempel, Nagel, Popper, Kuhn, Lakatos, Hesse, Grünbaum, Salmon and Laudan, uphold versions of standard empiricism as here characterized, in that there is general agreement that science does not, and ought not to, make any permanent metaphysical presuppositions about the nature of the world.

There are, I shall now argue, at least nine lethal objections to even this excessively modest version of standard empiricism.

1   Standard empiricism fails to solve the *practical problem of induction* (as it may be called), the problem, that is, as to how or why it can be rational to accept empirically verified or corroborated laws and theories of physics as a basis for *action*, via technological applications. Any law or theory of physics applies to infinitely many different empirical circumstances, but can only ever be 'verified' for finitely many of these. Thus, however much evidence is amassed in support of a law or theory of physics, we must always remain infinitely far away from verifying it empirically. Its probability, relative to established evidence, must remain zero. There can only be zero probability that the next standard application of our theoretical physical knowledge, however well verified, to building a bridge, aeroplane, radio or whatever, will meet with success. All knowledge of physical laws and theories must remain irredeemably speculative. How, then, can it conceivably be rational to base our actions on such improbable speculation?

It is worth noting that this practical problem of induction is only a part of the more general *problem of rational action* – the problem of characterizing, in general terms, what it is to act rationally, and of providing some kind of justification or rationale for the claim that action of this type does indeed deserve to be held rational.

It might be thought that standard empiricism only fails to solve the problem of induction insofar as rather strong claims are made for what science can achieve, namely that science can achieve theoretical knowledge sufficiently reliable to form a rational basis for action. Abandon entirely the claim that science can achieve such reliable knowledge, and – so it might be thought – there remains no problem of induction which cannot be solved within the framework of standard empiricism.

This is not the case. Even if no claim whatsoever is made about the capacity of science to achieve knowledge sufficiently trustworthy to form a rational basis for action, there is still a problem of induction remaining which standard empiricism cannot solve. Standard empiricism cannot, in other words, even explain how physics can acquire irredeemably *speculative* knowledge. This leads us to

2   Standard empiricism fails to solve the *theoretical problem of induction* (as it may be called), the problem, that is, of how and why it can be rational to accept empirically verified or corroborated laws and theories of physics as constituting merely the best available *conjectures or speculations* about the physical universe,

no claim being made to the effect that such speculations constitute reliable knowledge, a rational basis for action.

The problem of induction is often understood to be the problem of how we can know that the future will continue to resemble the past, the problem of how we can know that a theory verified in the past will continue to be verified in the future. It was roughly in this way that Hume formulated the problem. From our present point of view, however, there are several defects in this way of understanding the problem: the formulation of the problem needs to be improved.

In the first place we must remove any suggestion that the problem only arises if we presuppose an inductivist or verificationist conception of science, or presuppose that science can acquire reliable theoretical knowledge. Much more seriously than this, the *theoretical problem of induction* under consideration arises if we presuppose merely that it is rational to select theoretical *speculations* in science by means of observation and experiment alone. Not only inductivist and verificationist conceptions of science, but Popper's falsificationist conception of science too, all fail to solve the problem. In order to bring this out clearly, we can formulate the problem as follows.

Given any empirically successful physical theory T overwhelmingly corroborated in some inevitably finite space-time region R, we can easily construct endlessly many rival theories T_1, T_2, . . $T_{10}10$, . . , all of which agree with T in R but disagree with T elsewhere (in any way we please) at some other times and places. What rationale can there be, then, for preferring T (even as a mere speculation) on empirical grounds alone, since the theories T_1, T_2, . . $T_{10}10$, . . are all equally well corroborated by the available evidence?

T might be, for example, Newton's law of gravitation plus his laws of motion (presumed for the sake of the argument to be unrefuted). A typical T_1 would be:

Up to the end of today, an inverse square law holds, $F = \dfrac{G.M_1.M_2}{d^2}$; from tomorrow onwards, an inverse cube law holds, $F = \dfrac{G.M_1.M_2}{d^3}$.

Tomorrow, doubtless, T_1 will be refuted: endlessly many rivals to T will however remain unrefuted.

It deserves to be noted that the very absurdity of these endlessly many *aberrant* rivals to T (as they may be called) is a striking indication of the gravity of the problem confronting standard empiricism. If standard empiricism failed to provide a rationale for rejecting a fairly sensible rival to T, this would not constitute too serious an objection to standard empiricism. But the aberrant theories T_1, T_2, . . $T_{10}10$, . . are absurd; no physicist would take any such theory seriously for a moment. Thus any methodology of physics which fails to provide a rationale for rejecting such ludicrous theories, thereby fails disastrously.

But the problem confronting standard empiricism is even more serious than this would suggest. Possible physical systems, to which any sensible physical theory T applies, do not only differ with respect to position and time; they also differ with respect to values of other physical variables, such as mass, shape, density, velocity, temperature and so on. Thus, in order to develop endlessly many aberrant rivals to T, all just as highly corroborated as T, it is not necessary to find some region of space-time in which T has not been tested: all one need do is find some range of values of other physical variables (mass, temperature, etc.) for which T has not been tested, and arbitrarily modify the equations of T within this range of values in any way one pleases whatsoever. Thus if Newtonian theory has not been verified for physical systems consisting of bodies of density greater than some value D_0, or with relative velocities greater than some value V_0, or for bodies further apart than d_0, endlessly many highly corroborated aberrant rivals to Newtonian theory may be formulated, differing from Newtonian theory only in the as yet unobserved range of values of physical variables (density, velocity or distance). For example, one may stipulate that

$$F = G. \frac{M_1.M_2}{d^2} \text{ if } d < d_0, F = G. \frac{M_1.M_2}{d^r} \text{ if } d \geq d_0$$

where r is any number between, let us say, 1 and 4 but different from 2.

It should be noted that the infinitely many aberrant rivals to Newtonian theory, here indicated, do not in any way postulate (what may be regarded as) arbitrary changes of physical law at specific places or times. They all conform to the principle of uniformity of law in space and time; they are invariant with respect to space and time, or are 'strictly universal' in Popper's terminology (1959, pp. 62–70).

It is also important to note that there is no end to the ways in which physical systems, to which T applies, can be regarded as varying, and no end to the number of physical variables we can employ to distinguish different physical systems. These variables need not be referred to by T: it may simply be presupposed that variation of these variables leaves the applicability and success of T entirely unaffected. Thus the form of the equations of Newtonian theory not only remains invariant as we vary the place or time of a physical system, the mass, relative distance, velocity and acceleration of the bodies, their density and shape: the equations remain invariant as we change substance, temperature, colour, elasticity, smell. In order to formulate endlessly many empirically corroborated aberrant rivals to Newtonian theory, all we need do is specify, in universal terms (in terms of shape, substance, temperature, colour, smell or whatever) a kind of physical system, to which Newtonian theory applies, which has not yet been physically realized (perhaps because of its bizarre character): we then arbitrarily modify the Newtonian equations, in any way we please, for this specific kind of system. Thus we might stipulate: for two bodies, each of mass greater than two tons, each made of gold and shaped like a grand piano, adrift in space, an inverse cube gravitational law applies (but otherwise Newton's inverse square law applies).

Quite generally, given any theory T, in order to create endlessly many empirically equally successful, aberrant rivals to T, we need only take any kind of experiment E (however often repeated) and specify some bizarre, physically trivial but as yet untried physical modification of E, thus creating a nominally new kind of experiment E'. (The change from E to E' might involve such irrelevant modifications as painting the apparatus blue, placing an ounce lump of gold six feet from the apparatus, creating sound corresponding to middle C played on the violin.) Granted that T successfully predicts that E leads to outcome O, aberrant versions of T ($T_1, T_2, \ldots T_{10}10, \ldots$) agree with T everywhere except that for E' they predict O', where O' is whatever we may please. Empirically successful aberrant theories of this type are of course ludicrous, and would never be considered for a moment within science: all the more disastrous, then, is the failure of standard empiricism to provide a rationale for rejecting such ludicrous theories. (It is not physics that is here under attack, but rather a widely upheld, disastrously mistaken *philosophy of physics!*)

The problem confronting standard empiricism is still more serious than the argument so far indicates. Take any highly corroborated, sensible physical theory T: in practice T successfully predicts a range of phenomena A; it in principle applies to, but does not predict (because the equations of T cannot be solved) a range of phenomena B; it is ostensibly refuted by some recalcitrant, 'problematic' phenomena C; and it is entirely silent about a range of phenomena D (see figure 7). Here a phenomenon is understood, as it is in physics, to be a repeatable effect, a more or less observational or experimental law (which can of course always be sub-divided into as many sub-laws as we please). Consider now the rival theory T′ which asserts: in A, everything occurs as T predicts; in B, C, and D everything occurs in accordance with the experimentally established laws L_B, L_C, and L_D. T′ satisfies all the standard empiricist requirements demanded by Popper for a new theory to be more acceptable than its predecessor. T′ is not refuted whereas T is; T′ has far greater empirical content than T; T′ successfully predicts all the empirical success of T; T′ successfully predicts new phenomena, and given the infinite divisibility of laws, the infinite nominal variability of experiments, T′ can readily be shown to predict successfully hitherto 'unknown' phenomena. (And of course endlessly many aberrant alternatives to T′, equally preferable to T on empirical grounds alone, may also be constructed in the ways already indicated.)

In practice physicists persistently reject the most empirically successful (but horribly ugly) theories in favour of far less successful, and even ostensibly refuted, non-aberrant theories. The evidence persistently tells physicists that there are at most only patches of order in overall confusion, and physicists persist in believing in the existence of hidden order, despite all the evidence to the contrary, and even to the extent of rejecting empirically successful theories that postulate disorder. This scarcely accords with standard empiricism!

But the situation is even worse. It is not just that physicists persistently prefer beautiful, refuted theories to empirically far more successful, unrefuted but ugly theories: even worse, physicists habitually suppress evidence that clashes with established theoretical order. It is probably true to say that experimental results actually obtained in laboratories more often refute than corroborate established physical laws and theories: almost always experimentalists conclude that the experiment is at fault, and needs improving. It is nearly always extremely difficult, and a matter of great skill, to get apparatus used in even a fairly standard experiment to work properly. In practice, then, physical theory is used to refute experimental results at least as often as experimental results are used to corroborate or refute physical theory. In short, the conviction in physics that order exists in the world does not only lead to the rejection of empirically successful disorderly *theories*: it actually leads to the rejection of disorderly *evidence*.

3   Standard empiricism fails to solve the problem of providing a rationale for preferring simple to complex theories in physics.

One way in which one may seek to solve the above theoretical problem of induction is to stipulate that in physics simple, non-aberrant laws and theories are to be preferred to complex, aberrant laws and theories. Every proponent of standard empiricism acknowledges that considerations of simplicity play an important role in determining choice of theory in physics. What no proponent of standard empiricism can do, however, is explain how physics can select theories impartially and solely with respect to empirical success and failure if preference is persistently given to simple rather than complex theories (even to the extent, as we have seen, of the demand for simplicity persistently overriding the demand for empirical success). Persistently to prefer less empirically successful but simpler theories to more empirically successful but more complex theories is precisely to *abandon* the principle of empiricism, according to which theories are to be chosen solely with respect to empirical success and failure.

The only honourable attempt that I know of to solve this problem within the framework of standard empiricism is Popper's attempt as set out in *The Logic of Scientific Discovery* (1959, ch. 7). Popper's argument might be reformulated as follows. If theories are to be selected in science solely with respect to empirical success and failure, in the best, most honest and rigorous way possible, then preference needs to be given, other things being equal, to those theories most amenable, most sensitive, to being selected in this way – that is, to those theories that are the most *falsifiable* empirically, or, in other words, to those that have the highest empirical content. *But high falsifiability (or high empirical content) equals high simplicity.* The more falsifiable a theory is, so the simpler it is (and *vice versa*). Thus preference is given in science to simple rather than complex theories precisely in order to put impartial empiricism into practice in the most honest, rigorous way possible. Superficially, giving persistent preference to simple over complex theories violates the principle that only empirical considerations must be permitted to determine the choice of theory: actually we can only honestly put impartial empiricism into practice by persistently preferring simple (i.e. highly falsifiable) theories.

This argument requires, quite essentially, that whenever we increase the empirical content or falsifiability of a theory, we increase its simplicity. But this crucial thesis is false. As we have seen above, it is all too easy to increase the empirical content (or falsifiability) of a theory and at the same time vastly *decrease* its simplicity, vastly increase its complexity, its degree of aberrance (see figure 7).

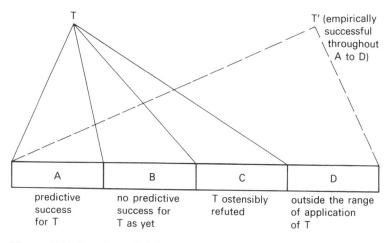

Figure 7 Refutation of claim that simplicity can be identified with empirical content

Subsequently Popper has in effect recognized the inadequacy of the theory of simplicity expounded in *The Logic of Scientific Discovery*. For in *Conjectures and Refutations* he puts forward a requirement of simplicity for science that is wholly in addition to falsifiability. He argues that a new theory, in order to be acceptable must 'proceed from some *simple new, and powerful, unifying idea* about some connection or relation (such as gravitational attraction) between hitherto unconnected things (such as planets and apples) or facts (such as inertial and gravitational mass) or new "theoretical entities" (such as field and particles) (1963, p. 241).

Granted that it can be formulated sufficiently precisely, this new '*requirement of simplicity*' (as Popper himself calls it) is in a sense more adequate than the earlier one in that it does perhaps exclude empirically successful, highly falsifiable aberrant theories of the kind discussed above. But how can selecting theories in accordance with the new requirement of simplicity conceivably be compatible with standard empiricism, with the principle that *empirical considerations alone* are to govern the choice of theory? If this new requirement of simplicity is adopted in physics, then potential new aberrant theories which clash with the requirement must be rejected whatever their empirical success might be. This is precisely to abandon standard empiricism. Even if the physical universe is complex and aberrant, this fact cannot be discovered, at the theoretical level, by a science that puts Popper's new requirement of simplicity into practice, just because any theory asserting aberrance in the physical world would be rejected out of hand, whatever its empirical success might be.

In short, a standard empiricist rationale can at most only be provided for Popper's earlier, wholly inadequate requirement of simplicity (explicated in terms of falsifiability). As far as Popper's later, and perhaps more adequate, requirement of simplicity is concerned (which cannot be explicated in terms of falsifiability), no rationale can be provided for adopting the requirement within the framework of standard empiricism.

4   Quite apart from being unable to provide a rationale for adopting a principle of simplicity in physics, standard empiricism fails even to specify adequately what simplicity is. Simplicity cannot be defined merely in terms of the number of postulates a theory has, since number of postulates can always be reduced as much as we please, by logical means, down to one postulate. Nor can simplicity be defined in terms of number of different sorts of entities postulated by a theory, since number of different sorts of entities can always be artificially reduced by stipulating that different sorts of entities are different states of *one* entity. For similar reasons, simplicity cannot be defined in terms of the number of different sorts of basic physical properties the theory attributes to physical entities. Simplicity cannot be defined merely in terms of the mathematical simplicity of the equations used to formulate the theory, since this depends on our choice of mathematical and conceptual conventions, a suitable change of terminology and concepts being sufficient to transform the 'simplest' theory (in any such sense) into a highly complex theory (and *vice versa*). There are, it is true, standard empiricist attempts such as Goodman's (1972, ch. 7) to give a precise, formal explication of the notion of simplicity: these seem to have nothing to do with physics. In the circumstances it seems more honest for standard empiricists to confess, with Popper, that the '*requirement*

of simplicity is a bit vague, and it seems difficult to formulate it very clearly' (1963, p. 241).

In order to make clearer what simplicity is in the context of physics, one crucial problem must be solved, namely the problem of distinguishing sharply mere *terminological simplicity*, of no account in assessing the acceptability of a theory, from *physical simplicity*, highly relevant to the assessment of a physical theory. Standard empiricism fails to solve this problem for at least one very good reason. Adoption of a *requirement of physical simplicity* in physics (physical simplicity here being clearly distinguished from mere terminological simplicity) would clearly and explicitly violate standard empiricism. In other words, clarification of the methodologically important notion of simplicity, as this arises within physics, can only be done if standard empiricism is rejected.

5   Not only does standard empiricism fail to provide a rationale for the acceptance and rejection of theories in the light of experimental results: it fails even to provide a rationale for the acceptance of experimental results themselves. As far as physics is concerned, an experimental result is a repeatable effect, a physical law. Any particular observational or experimental result, obtained at a particular time and place, can only become a part of physical knowledge, capable of corroborating or refuting physical theory, insofar as the particular result is construed to exemplify a repeatable, universal, law-like effect. Inevitably, however many experiments are performed, we must remain infinitely far away from verifying any such law-like experimental phenomenon. The two problems of induction, discussed above in connection with the acceptance of *theory* in physics, arise just as potently in connection with the acceptance of *experimental results*, as these are understood in physics.

6   Standard empiricism fails to specify clearly and unambiguously the *methods* of physics – quite apart from its failure to provide a rationale for any such methods. Thus proponents of standard empiricism, such as Hempel, Carnap, Nagel, Popper, Kuhn, Lakatos, Grünbaum, Hesse and Salmon, disagree substantially about the nature of scientific method. Again, no clear, unambiguous formulation of the methodologically vital 'requirement of simplicity' is forthcoming, as we have seen.

7   Views about the aims and achievements of physics, the nature of physical theory, traditionally associated with standard empiricist

philosophies of physics, fail to solve the problem of the *miraculous predictive success of physical theory* (as it may be called). According to these traditional standard empiricist philosophies, physical laws and theories essentially do no more than assert the existence of *regularities* in phenomena. At once the problem arises: why should phenomena observe these postulated regularities? A vast range and diversity of phenomena obey the regularities postulated by 'well-established' physical theory, to a quite extraordinary degree of accuracy. Unless something exists in the world which is, in some sense, 'responsible' for this extraordinarily widespread, accurate observance of postulated regularities – which in some sense 'controls' or 'determines' phenomena to obey these regularities – the continuing predictive success of physical theory can only be deemed to be utterly incredible, an enduring miracle. As traditionally interpreted, there is no scope for physical theory to assert the existence of anything in the world capable of being 'responsible for', in any sense, lawfulness or regularity. According to these traditional views, observed regularities can only be 'explained' by the discovery that they are a part of (and can be approximately derived from) even more widespread and accurate regularities (as when the regularities of Kepler's laws of planetary motion are discovered to be a part of the more universal regularities of Newtonian theory). This kind of 'explanation' only deepens the mystery: it serves only to make it all the more incomprehensible why phenomena should comply with such universal regularities to such an incredible degree of accuracy.

8   Standard empiricism fails to explain how it can be possible for theoretical physics to make progress. In attempting to discover new, better theories, physicists are, according to standard empiricism (as we have seen) confronted by infinitely many possible theories. The likelihood of formulating a theory that constitutes an improvement over existing theoretical knowledge would thus seem to be infinitely remote. Standard empiricism cannot provide even a fallible rational guide for the development of good, new, physical theories, since according to standard empiricism the only rational criteria that exist in physics for the assessment of ideas have to do with empirical success and failure of theories once they have been formulated. Theoretical physicists cannot even restrict their attention to new theories that are compatible with existing well-corroborated theories, since more often than not successful new theories are *incompatible* with pre-existing theories. (Thus quantum theory is incompatible with classical physics, Einstein's

general theory of relativity is incompatible with his special theory of relativity, in turn incompatible with Newtonian theory, which is in turn incompatible with Kepler's laws of planetary motion.)

9   Not only does standard empiricism fail to provide a rationale for the claim that science makes progress (objections 2 and 3); not only does it fail clearly to characterize methods designed to achieve scientific progress (objection 6); not only does it apparently render the achievement of scientific progress little short of the miraculous (objections 7 and 8); but wholly in addition to all this, there is, if anything, an even more devastating failure: within the framework of standard empiricism there is no solution to the problem of what scientific progress *means*.

As long as science progresses by accumulating more and more truth, no problem arises as to what progress means. We have just seen, however, that science does not progress like this: new theories tend to correct their predecessors, thus revealing these predecessors to be, strictly, false. Physics in particular develops from one false theory to another: it is this which creates the problem of what scientific progress means.

The only conceivable solution to this problem, within the framework of standard empiricism, would seem to be the solution proposed by Popper: physics progresses if and only if successive theories T_1, T_2, T_3 . . ., though all false, nevertheless get progressively closer and closer to truth, in that they have progressively more and more truth in them, and/or less and less falsehood. A little more precisely, given any two false theories, T_1 and T_2, T_2 is closer to the truth than T_1 if and only if *either* (a) T_1 and T_2 have precisely the same false consequences but the true consequences of T_1 are less than, in the sense of being a proper part of, the true consequences of T_2; *or* (b) T_1 and T_2 have precisely the same true consequences but the false consequences of T_2 are less than, in the sense of being a proper part of, the false consequences of T_1; *or* (c) the true consequences of T_1 are less than those of T_2, and the false consequences of T_1 are greater than those of T_2 (case (c) being, as it were, an amalgamation of cases (a) and (b)) (Popper, 1963, pp. 231–7).

It turns out, unfortunately, that these apparently unproblematic conditions can never be realized for any false theories T_1 and T_2 (Tichý, 1974; Harris, 1974; Miller, 1974). If T_2 has more true consequences than T_1, then it inevitably also has some false consequences in addition to those of T_1 (so case (a) cannot be realized). Alternatively, if T_1 has more false consequences than

T_2, then it inevitably also has some true consequences in addition to those of T_2 (so case (b) cannot be realized). As neither case (a) nor case (b) can be realized, case (c) cannot be realized either.

For case (a), consider propositions p and q such that p is false, and is implied by T_1 (and T_2), and q is true, is implied by T_2 but not by T_1. In this case the proposition 'p & q' is false, is implied by T_2, and yet is not implied by T_1. Hence T_2 has false consequences that are in addition to those of T_1, and case (a) cannot be realized. In connection with case (b), let r be a proposition that is false, that is implied by T_1 but not by T_2; let s be a proposition that is false and is implied by both T_1 and T_2. In this case the proposition 's⊃r' is true, is implied by T_1 but is not implied by T_2. Hence T_1 has true consequences that are in addition to those of T_2, and case (b) cannot be realized. Hence case (c) cannot be realized either, it thus being impossible to compare any two false theories with respect to their closeness to the truth. (See figure 8 for a representation of this essentially simple argument.)

If some kind of distinction can be made between atomic propositions (p, q, etc.) and molecular propositions (p & q, s ⊃ r, etc.), then it does become possible to compare some false theories with respect to their closeness to truth, in that for the purposes of the comparison, molecular propositions, the source of the problem,

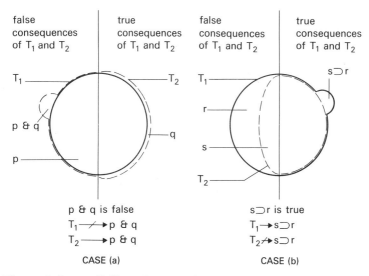

Figure 8 Impossibility of comparing verisimilitude of two false theories

can be excluded from consideration. However, any distinction between atomic and molecular propositions must be relative to our choice of language. This has the consequence that the question of which of two false theories is closer to truth can depend on our choice of language. In choosing a language, a conceptual scheme, we may in effect be deciding or determining what is for us significant, important, of value. Thus the question of which of two false theories is closer to the truth can depend on decisions of significance and value. In no way damaging to aim-oriented empiricism and the philosophy of wisdom, this conclusion is a disaster for standard empiricism.

I conclude, in view of the above nine objections, that even the extremely modest version of standard empiricism formulated above is untenable, and must be rejected.

The mistake of standard empiricism is to misrepresent the basic intellectual aim of physics. The aim of physics is not, and cannot be, merely to improve knowledge about the world, nothing being presupposed about what sort of world this is. On the contrary, in the contexts both of verification and discovery, an essential aim of physics is to improve our understanding of the world, it being an unavoidable presupposition of physics that understanding is in principle possible, the universe being, in some way or another, comprehensible. There are many different sorts of ways in which the universe might be comprehensible. Modern physics, a highly sophisticated and successful development of humanity's long-lasting search for understanding, presupposes, more or less specifically, that the universe is intelligible in the sense that some kind of unified pattern runs through all phenomena. The vital general point, however, is that the pursuit of knowledge cannot be dissociated from the pursuit of understanding – from the presupposition that the world is such that understanding of some kind or another is in principle possible. At first sight it might seem that the pursuit of knowledge can be, and ought to be, dissociated from any presupposition about the world whatsoever, including the presupposition that the world is comprehensible: it is just this standard empiricist thesis that has been shown to be untenable in the last section.

What is meant by the assertion that the universe is comprehensible? I wish to allow that there are many different ways in which the universe may be comprehensible and thus, in a sense, many different possible more or less specific conceptions as to what comprehensibility is. Among the many ways in which the universe might conceivably be comprehensible, there are two – here called *personalism* and *physicalism* – which are of particular importance in the history of human thought.

According to personalism (sometimes called animism), the world is made up of beings – or gods – somewhat like persons, with purposes, desires, feelings, experiences, acting in response to experiences in order to satisfy desires, realize intentions. What goes on in the non-human world is the outcome of the actions of these gods rather as what goes on in human society is the outcome of the actions of people. Such things as the sun, the moon, the sea, the earth, forest, mountain, river, sky are, for versions of personalism, different gods, the characteristics and behaviour of these things being the expression of the intentions of these gods rather as the behaviour of a person is normally an expression of that person's intentions. Other versions of personalism hold that there is but one God, the whole world being the expression, the outcome, of God's will.

Physicalism, in sharp contrast to personalism, holds that the world is entirely impersonal in character. According to physicalism, although there appear to be very many different sorts of things and phenomena in the world, changing and interacting in apparently very many different and often arbitrary ways, in reality the world is made up of only a very few different sorts of things (atoms, point-particles, fields or whatever) which change and interact in only one, precise, fixed way. That which does not change, X, precisely determines the manner in which that which changes, Y, does change (both X and Y being properties of the basic physical entities out of which everything is composed). In the case of atomism, for example, X is the unchanging properties of space, time and atoms, whereas Y is the relative positions and motions of the atoms at any given moment.

There are other ways in which the universe may be conceived to be comprehensible. Personalism may be regarded as a special case of a more general view – *purposivism* – according to which the characteristics and behaviour of things can be understood in terms of purposes they realize (specific purposes contributing rationally perhaps to some overall cosmic purpose), these purposes not necessarily being conscious purposes, the purposes of person-like gods, or one God. Some versions of personalism accommodate aspects of physicalism – versions which hold that the material world behaves in accordance with physicalism but has been created to do so by God. Other versions of personalism may conceive of the material world as having been created by God to express thought and feeling, somewhat like a book or a work of art, the world thus being imbued with, and explicable in terms of, Divine meaning or beauty. Alternatively, versions of personalism may conceive of God's intentions in creating the world more functionally, on analogy with a builder or architect.

These diverse conceptions of comprehensibility, despite their obvious differences, do have some things in common. They all agree that the comprehensibility of the universe has to do with its overall character, organization or structure. Each conception of comprehensibility takes some familiar, small-scale example of understanding – understanding a person, a persisting object or regularity, an artefact, a book or work of art, a pattern or machine – and projects this small-scale case of comprehensibility onto the entire universe. The character or organization of parts or bits of the world are taken to be indicative of the character or organization of the whole of the universe. Finally, comprehensibility may be held to be linked to *knowability*. To the extent that something is incomprehensible to us, we cannot readily acquire knowledge about it: to the extent that we do understand it, we can readily improve our detailed knowledge of it as needed. If we have a good overall understanding of a person, we can quickly come to know what the person is feeling, thinking, intending, doing, on particular occasions: if we lack such understanding, our ability to acquire such knowledge is much more limited and fallible. The diverse classic conceptions of comprehensibility may thus be said to take some familiar small-scale case where knowledge is successfully progressively acquired and improved owing to the existence of some kind of persisting order or structure; this small-scale order or structure is then projected onto the whole world.

It is, I suggest, in the light of these considerations, reasonable to hold that there is a general notion of comprehensibility, common to all the diverse, more specific notions. Quite generally, to say that the universe is (in some way or other) comprehensible – as the term is used here – is to assert that there is some kind of overall uniformity, lawfulness, order, coherence, pattern, meaning or plan to the universe, in terms of which particular things and events can in principle in some way be explained and understood. It is to deny mere disorder, incoherence, arbitrariness or randomness. It is to assert that characteristic reasons or explanations exist (even if not known) for the way things are, for the way things change (in terms of the underlying order or intelligibility pervading the world). To assert that the universe is in some way or another comprehensible is to assert that there exists something – the characteristic order or meaning of the universe – which pervades all that there is. In conceiving of a possible universe that is in some way comprehensible we at the same time, implicitly or explicitly, conceive of a vast number – usually infinitely many – possible universes that are not comprehensible in this way, in that in them the characteristic kind of order or pattern involved does not pervade all that there is. In a comprehensible universe, in other words, all merely *aberrant* theories, of the kind discussed in the previous section, are false. To say that the universe is comprehensible is to make a substantial (but probably untestable or metaphysical) assertion about the overall character or nature of the universe: the more specific and precise the sense in which the universe is asserted to be 'comprehensible', so the more substantial and precise the assertion becomes (and *vice versa*).

Comprehensibility is such, so we may hold, that in a comprehensible universe any fragment of the universe contains clues or guidelines as to the character of the whole of the universe – somewhat as a small piece of patterned wallpaper contains clues as to the whole wallpaper. Any fragment of a comprehensible universe will exemplify the characteristic kind of uniformity, order, pattern or meaning that pervades the whole: it is this, exemplified in any fragment, which provides clues or guidelines to the nature of the rest of the universe.

I turn now to a consideration of the all important questions: to what extent is it true that the comprehensibility of the universe, in the sense indicated, is a necessary and sufficient condition for the universe to be *knowable*, for it to be rationally possible at least for genuine knowledge to exist and grow? Does this provide an adequate rationale for committing physics to the presupposition that the universe is comprehensible? Or can such a rationale be provided in some other way?

To begin with, it seems reasonable to hold the following. In a universe that is in some way comprehensible, in the sense indicated, there is at least a rational hope or possibility – as opposed to an infinite improbability – that knowledge can be improved. For, so we may argue, the uniformity, order, pattern, meaning or plan that pervades the whole of a comprehensible universe makes it rationally possible that even *imperfect* knowledge of a *part* of the universe can become more perfect knowledge of the whole universe. Since in any fragment of a comprehensible universe there exist clues or guidelines as to the nature of the rest of the universe, it is always reasonable to suppose that imperfect knowledge of such a fragment will itself contain, explicitly or implicitly, clues or guidelines as to the nature of the rest of the universe – clues or guidelines which make possible the progressive extension and improvement of knowledge. In such a universe, in other words, there can exist a body of knowledge which contains, explicitly or implicitly, heuristic and methodological rules (reflecting, perhaps imperfectly, the comprehensible character of the universe) designed to make possible, to promote, the improvement of knowledge.

On the other hand, we may also argue that in a wholly incomprehensible universe, knowledge cannot grow, or even exist. In a wholly chaotic universe, devoid of any kind of uniformity, order, pattern or meaning even to a limited extent in even a limited region, there cannot be persons, language, life, goal-pursuing or knowledge of any kind. For all these things exemplify and thus require for their existence at least some limited degree of order, pattern or meaning of some kind or other: and just this has been denied.

In a comprehensible universe, the growth of knowledge is a rational possibility: in a wholly incomprehensible universe, knowledge cannot even exist, let alone grow. These considerations rationally entitle us to commit science to the presupposition that the universe is in some way comprehensible, to some extent at least. Ultimately all our knowledge is conjectural in character. The assumption that the universe is comprehensible to some extent is no exception: it is a conjecture which we cannot prove to be true, or conclusively verify. On the other hand we can be quite certain that the opposite assumption 'The universe is wholly incomprehensible' cannot become a part of scientific knowledge, since if this is true there can be no science and no knowledge. We thus risk nothing in holding that 'The universe is comprehensible to some extent' is a solid, undeniable item of scientific knowledge. If this proposition can be asserted, it cannot be wrong.

This argument does not establish, however, that we are rationally entitled to commit physics to the presupposition that the universe is *wholly* comprehensible. It is logically possible that we might exist and possess knowledge, and science might meet with success, even though only a *part* of the universe is only *approximately* comprehensible. There are at least four general possibilities to consider. (1) Our space-time region of the universe is comprehensible, but elsewhere incomprehensible phenomena occur. (2) Comprehensibility extends throughout all regions of space and time, but is intermittently violated in that occasional incomprehensible, miraculous events occur. (3) Comprehensibility is restricted to the kind of phenomena that can be produced or observed on or from the earth: there are extreme phenomena that lie outside this range (at high energies, for example, or high densities of matter) that are incomprehensible. (4) The universe is only approximately or asymptotically comprehensible, in the sense that it is such that theoretical physics must always be infinitely far from its goal, infinitely many revolutionary theoretical developments being needed before a unified theory could be formulated which specifies precisely how the universe is comprehensible. Now, on what grounds do we reject these and accept instead the following? (5) The universe is in some way wholly and precisely comprehensible, without exceptions. It has been established that some degree of comprehensibility must be presupposed to exist by science. What rationally entitles us to commit science to the complete comprehensibility of (5), as opposed to the partial comprehensibility of (1), (2), (3) or (4)?

In answering these questions, one important preliminary point needs to be appreciated. On the face of it, it might seem more modest, more cautious, and therefore more rational, for physics to assume no more than partial and approximate comprehensibility, rather than (some unknown kind of) complete and precise comprehensibility, pervading all that there is. This impression of greater modesty is, however, somewhat illusory. In assuming only that the universe is at least in part approximately comprehensible, we thereby assume that the rest of what there is, even if not wholly comprehensible, nevertheless is sufficiently stable and well-behaved not to disrupt the partial comprehensibility in existence. Not just science, but even our most trivial, common-sense claims to knowledge contain, explicitly or implicitly, presuppositions about the entire cosmos, to the effect that all that exists of which we are ignorant is sufficiently stable and well-behaved not to disrupt the small bit of the universe we claim to know and understand. Insofar as we know anything about anything (and we cannot live or do science unless we possess some knowledge), we also know something at least about the entire cosmos, all that there is. And the more extensive and precise our scientific claims to knowledge are (however conjectural), so the more restricting and precise must be our assumptions about the character of all that exists of which we are largely ignorant.

What needs to be established, then, is that it is more rational for physics to presuppose, as conjectural knowledge, that the universe is, in some way, wholly and prefectly comprehensible, rather than only partly, approximately comprehensible, and elsewhere at least sufficiently stable to permit approximate comprehensibility to persist.

Here is my argument in support of this contention.

The refutation of standard empiricism above showed conclusively that physics, in order to progress, in order to develop (conjectural) theoretical knowledge, must persistently reject empirically successful, *aberrant* theories. This persistent rejection of empirically successful, aberrant theories means that physics, implicitly or explicitly, presupposes that the world is non-aberrant – that it is, in some way or other, comprehensible. It is possible to pursue physics in such a way that this presupposition of non-aberrance or comprehensibility is left as a vague, unassertive, implicit, even unacknowledged, but nevertheless profoundly influential assumption. Physics pursued in this way is, however, seriously irrational, seriously lacking in intellectual rigour, just because a substantial, influential and profoundly problematic assumption is not explicitly formulated and explicitly criticized. Quite generally, *an essential requirement for rationality, for intellectual rigour, is that substantial, influential and problematic assumptions be made explicit so that they are open to criticism and thus, one may hope, improvement.* Thus, if physics is to be genuinely rigorous and rational, it is essential that the substantial, influential and profoundly problematic assumption of non-aberrance or comprehensibility be formulated explicitly in as clear, bold, decisive, precise, extensive, extreme and contentful a way as possible, so that it may be subjected to the maximum degree of sustained criticism, in this way, we may hope, becoming progressively improved. Our best scientific conjecture as to how the universe is perfectly comprehensible, put forward at any stage in the development of science, is the tentative spearhead of research into the unknown, the probing searchlight we shine into the darkness ahead in the hope of lighting up our way. A conjecture postulating perfect, precise comprehensibility, as opposed only to partial, approximate comprehensibility, is to be preferred – is more rationally acceptable – (other things being equal) because it offers more, because it is potentially more helpful for the progress of physics, and because it is more *vulnerable to criticism*, more open to being found wrong (in ways to be discussed) should the universe be comprehensible in some other way.

In pursuing physics we are rationally entitled to adopt any (morally acceptable) procedure which can only help, and cannot hinder, scientific progress, the acquisition of genuine knowledge and understanding. Consider now the following procedure: formulate the best possible conjecture as to how the universe is perfectly and precisely comprehensible, in as explicit and definite a form as possible, in the light of all that is known and understood, this conjecture being *both* persistently criticized, *and* employed to assess the acceptability of empirically successful physical theories. Adopting this procedure, I claim, can only help, and cannot hinder, the progress of physics, whether the universe is in reality perfectly comprehensible, only partially comprehensible, or wholly incomprehensible. For if the universe is perfectly and precisely comprehensible, the procedure indicated can clearly only aid the progress of physics. If the universe is only partially comprehensible, the procedure indicated can still only aid, and cannot hinder, the progress of physics for in this case, in order to improve knowledge and understanding, we can do no better than to seek for perfect comprehensibility, and fail. To assume only partial comprehensibility at any stage cannot help, and may well hinder progress in that there may exist more comprehensibility (and therefore more potential growth of knowledge) than we have allowed. We can only discover that the universe is only partially comprehensible (to the extent that this can be discovered) by persistently seeking for perfect comprehensibility, and failing in the attempt. Finally, if the universe is wholly incomprehensible, there can be no life, no knowledge and no science: thus even in this case, the procedure indicated cannot hinder scientific progress (since in this case there can be no science at all). Therefore in pursuing physics we are rationally entitled to adopt the procedure indicated.

I conclude that for physics to be intellectually rigorous and rational it is essential that physics explicitly formulates and criticizes, as a part of scientific knowledge, the best possible conjecture, at any stage, as to how the universe is wholly and precisely comprehensible, in the way just indicated. Furthermore, whenever there is some acceptable notion of degrees of comprehensibility, physics is entitled to assume that the universe is comprehensible to the highest degree, other things being equal.[1]

This argument, this proposed solution to the problem of induction, the traditional problem of the rationality of science,

---

[1]Some further aspects of this proposed solution to the traditional problem of the rationality of science can be found in Maxwell (1972a, pp. 131–52; 1974, pp. 123–53 and 247–95; 1979, pp. 629–53). Others have advocated conjectural, presuppositional approaches to solving the problem of induction. See, for example, Russell (1948, pt. VI); Burks (1977, ch. 10). My criticism of these attempts is that they

might be said to be Kant's attempt at a solution radically improved in the light of Popper's or, alternatively, Popper's attempt at a solution improved somewhat in the light of Kant's. The argument takes seriously Kant's central point that in order to solve the problem of knowledge, highlighted especially by Hume, we need to take into account what must be the case if knowledge is to exist at all, the necessary conditions for knowledge to be possible. But it also accepts Popper's point that all our knowledge is ultimately irredeemably conjectural in character. The metaphysical theory 'The universe is wholly and perfectly comprehensible' is the only proposition that might remotely exemplify Kant's key notion of a 'synthetic *a priori*' proposition – a proposition about the world known independently of experience. However, it exemplifies Kant's notion in only the weakest possible sense. The proposition is wholly conjectural in character (and not something that can be known to be true with certainty, as Kant would insist). Insofar as we have a good reason to accept this proposition as a part of our conjectural scientific knowledge *on grounds that are independent of experience*, this can only be done if the proposition is interpreted in the loosest, vaguest way conceivable – 'comprehensible' having only its very general meaning indicated above. In order to have good reasons to accept a more precise version of the proposition or theory – physicalism, for example, or some version of physicalism such as atomism – it is essential to appeal to experience, to the actual development of empirical science. (Thus nothing as precise as 'Space is Euclidean' can be a synthetic *a priori* proposition in even the weakest sense.) The above argument also accepts Popper's point that the problem of knowledge is fundamentally the problem of the *growth* of knowledge. Actually even the mere *possession* of knowledge – scientific or common-sense – involves in a sense the acquisition of new knowledge, as time passes, and things persist and change. Even for animals, the key problem is the acquisition or growth of knowledge, the capacity to detect, for example, the presence of other animals from the minutest of signs or evidence, whether these animals be predators, food, rivals, a mate or offspring. Thus Kant's idea is transformed into the idea of specifying the best possible conditions that we are rationally entitled to postulate in order to render the

---

attempt to justify as rational what is actually irrational, namely that science should make a too narrowly conceived, fixed metaphysical presupposition about the nature of the world, and should adopt fixed methods. They fail to characterize science as learning in what way the universe is comprehensible as it proceeds. They fail, in other words, to advocate aim-oriented empiricism.

growth of knowledge as rapid as possible. In addition, the above argument incorporates Popper's point that in order to solve problems of knowledge created by the sceptical arguments of Hume and others, we must forego the attempt to specify some restricted body of knowledge that is wholly immune to scepticism, that is absolutely certain, beyond all doubt. Instead of seeking to defend reason and knowledge by attempting to *refute* scepticism in this way, we need to recognize that scepticism is absolutely essential to reason and the growth of knowledge. It is precisely by exposing our claims to knowledge to ruthless scepticism, to persistent, savage criticism, that we can best hope to make progress. The endeavour to delimit, defuse or rebut scepticism is actually profoundly *irrationalist*, profoundly damaging to the growth of knowledge. Nothing is immune to doubt – certainly not the theory that the universe is in some way or other comprehensible. We need, indeed, to be sceptical even of scepticism itself, of its capacity invariably to aid the growth of knowledge. But we must not be totally (uncritically) sceptical about scepticism, its value, its capacity to aid the growth of knowledge and understanding, or there is a danger that scepticism will destroy itself and lapse into its opposite, dogmatism. If one doubts the value of scepticism to the point that one doubts that it is possible by means of criticism to improve knowledge at all, to assess correctly the relative merits of rival claims to knowledge, there is always the danger that one will end up accepting some ostensible body of knowledge as being as good as any other, criticism being ignored on the grounds that all criticism is ultimately pointless. In practice this is extreme dogmatism. Extreme scepticism of the value of scepticism is thus to be resisted as self-destructive and counterproductive. And more generally, whenever it can be shown that any application of doubt or scepticism can only hinder, and cannot aid, the growth of knowledge and understanding, then we are rationally entitled to abstain from this kind of doubt, on pragmatic grounds. (In this way we are *rationally*, and not just destructively, sceptical of scepticism.) Popper himself employs this kind of pragmatic argument when, for example, he argues that we ought not to doubt all our knowledge all at once, since this sabotages the very possibility of improving knowledge (1963, pp. 238–9). The argument I have outlined above is an example of just such a pragmatic, rational, critical delimitation of scepticism, in that the argument is designed to establish that to doubt that the world is in some way comprehensible cannot help, and can only hinder, the growth of knowledge. Again, the argument outlined above accepts

Popper's contention that we need to put forward and prefer (other things being equal) those ideas, those theories, which make the boldest, the most ambitious, substantial and extensive claims about the world, because in doing this we give ourselves the best chances of discovering error and making progress. The argument accepts, that is, Popper's contention that – as one might put it – scientific rigour involves being constrained to accept (other things being equal) our wildest, furthest flung imaginings. It is Popper's hope that this principle suffices to solve the problem of selecting the best theory in the light of evidence, granted that there will always be infinitely many rival theories compatible with all available evidence to choose from. According to Popper, we choose the most ambitious theory, the theory which asserts the most, is the most falsifiable. We saw above that this proposal fails. It fails because it is too timid, not bold enough. With even greater daring than Popper envisages, we need to conjecture that the entire cosmos is in some way perfectly and precisely comprehensible, this conjecture being upheld with almost reckless audacity throughout all the ups and downs of empirical research – even though, of course, every empirical hint as to the particular way in which the universe may be comprehensible is eagerly seized on and exploited. Invariably we choose those empirically successful theories which accord best with our best conjecture as to how the universe is comprehensible. We are rationally entitled to do this just because it gives us our best hope of improving our knowledge and understanding of the world. Confronted by infinitely many possible universes, we can do no better than explore first all the perfectly comprehensible universes that we can imagine: only when we have exhausted all our ideas about comprehensible universes will there be any point in considering those that are partially, approximately comprehensible. Finally, and most important of all, the above argument accepts, and arises as the outcome of putting into practice, Popper's key point that *criticism* lies close to the heart of *rationality*. Physics pursued in accordance with some version of standard empiricism such as Popper's falsificationism, in selecting from infinitely many empirically successful theories only those few that are non-aberrant, must inevitably, implicitly, be accepting a massive, profoundly influential metaphysical conjecture about the world, to the effect that the world is non-aberrant. This metaphysical conjecture is built into the actual (as opposed to the declared) methodology of physics – into the actual practice of rejecting empirically successful aberrant theories out of hand. This massive and profoundly influential conjecture about the nature of the world cannot however be criticized explicitly within science just because of the metaphysical (or untestable) character of the conjecture. The intellectual standards of standard empiricism, and especially falsificationism, are such that untestable, metaphysical conjectures are to be excluded from science. Aim-oriented empiricism, by contrast, insists that our best more or less specific conjecture as to how the universe is comprehensible or non-aberrant, in the light of present knowledge and methods, must be formulated explicitly as a vital part of scientific knowledge, precisely so that it can be criticized, and thus, we may hope, improved. Physics pursued in accordance with aim-oriented empiricism is thus more rigorous, more rational, than it is when pursued in accordance with any version of standard empiricism, including falsificationism, just because aim-oriented empiricism insists on, and standard empiricism prohibits, sustained, explicit criticism of a massive conjecture about the world inevitably exercising a profound influence over scientific thought. Furthermore, as we shall see, this enhancement of intellectual rigour, of scientific rationality, has important consequences for the way we do, teach and understand science. Adopting aim-oriented empiricism encourages, and adopting standard empiricism discourages, improvement of aims and methods as science proceeds – vital for scientific progress.

In short, if the solution to the problem of induction offered here faintly echoes Kant's, it does so only by being even more radically Popperian than Popper's own attempt at a solution.

It is perhaps just worth noting that there are two additional ways in which the solution to the problem of induction offered here differs from Kant's. According to the solution offered here, it is ultimate reality, the world as it is in itself, remote from human experience – Kant's *noumenal* world – that we are rationally entitled to claim to know to be in some way comprehensible, in an *a priori* but conjectural manner. For Kant, all knowledge, including all *a priori* knowledge, is only about the *phenomenal* world, the world of possible experience, it not being possible to know anything about noumena except that they exist. The solution offered here thus rejects Kant's 'Copernican revolution'. The world is not comprehensible to us because what we call the world conforms to our minds: rather, it is only comprehensible to us if we allow it to teach us in what way it is to be understood. Instead of our mind shaping the (phenomenal) world to conform to our ideal of comprehensibility, we must seek to shape our ideal of comprehensibility to conform to the character of the (noumenal) world. Kant's mind-centred anthropomorphism (actually wholly non-Copernican in spirit) is replaced by the anti-anthropomorphism of modern physics. As the physicist John Wheeler has emphasized, in order to understand the world, we must be prepared to recognize how unfamiliar, how mysterious it is.

It is perhaps also just worth noting that other, somewhat less Popperian arguments may be given in support of the contention that we are rationally entitled to hold as a part of scientific knowledge that the universe is wholly and not just partially, comprehensible. Modern physics is based, so we may argue, on the presupposition that the universe is impersonal in character, there thus being nothing privileged about humanity's position in the cosmos, as far as the ultimate nature of the cosmos is concerned. This anti-anthropomorphic presupposition requires us to hold that comprehensibility (not necessarily precisely as characterized by us at present) is not confined to our particular region of space-time or phenomena, but is present throughout all places, times and phenomena. Again, we may argue that physics quite properly seeks understanding as being of value in itself, and not merely as an aid to the acquisition of knowledge. Granted that we seek understanding, we are entirely justified, on pragmatic grounds, in holding that what we seek exists, since the only way we can discover it does not exist, is to search for it and fail to discover it. Postulates of partial comprehensibility are in this case even more obviously unhelpful. Finally, it may be argued that the possibility that the universe is only partially approximately comprehensible is to be rejected because of its wild implausibility. In any such universe, there can be no reason, no explanation, as to why limited chaos does not spread like a contagion to engulf all order. Once some inexplicable events occur, there can be no reason why all events should not be inexplicable. Sustained order could only be an infinitely improbable accident. Order of this kind, without any underlying *raison d'être*, is too absurd to deserve a moment's consideration.

What all this amounts to, then, is that physics in particular, and natural science more generally, insofar as they have sought knowledge for its own sake, have suffered from rationalistic neurosis (in the sense indicated in chapter 5). The actual aim of natural science, quite properly, is to improve our knowledge and understanding of a world presupposed to be comprehensible. The declared aim (as far as most scientists and philosophers of science are concerned) has been to improve knowledge about the world, no presupposition being made about the nature of the world (theories being selected, in the end, solely with respect to empirical success and failure). This declared aim and method of science, standard empiricism, being widely upheld, has exercised a profound influence over science itself. Many physicists, and scientists more generally, do of course declare a personal faith in the comprehensibility of the universe: this is however at most a personal conviction, upheld in the context of discovery (where everything is permitted). The decisive point is that modern science as a whole does not hold the metaphysical theory 'The universe is comprehensible' to be a vital part of public, objective scientific knowledge, in the context of verification. The declared aim, in the context of verification, is the discovery of *truth*, and not *explanatory truth* (it being mistakenly held that in this context the truth cannot be presupposed to be explanatory).

This long-enduring rationalistic neurosis of science has had all the usual damaging consequences mentioned in chapter 5. The more 'rationally' or honestly science pursues its declared aim, so the worse off it is: scientific success seems to require some measure of irrationality. The attempt to construe science as rational, given its declared aim, can only be counter-productive. Such an attempt seeks to solve counter-productive neurotic problems (the problems discussed on pages 110–17 above). This explains why so much orthodox philosophy of science, presupposing standard empiricism, is so unfruitful from the standpoint of science itself, and is treated so dismissively by so many scientists. All this helps to ensure that the neurosis persists since reason (or critical concern for basic aims and methods) becomes discredited. In moving from standard empiricism to aim-oriented empiricism we transform the neurotic, insoluble, and scientifically sterile problems discussed above into problems whose solutions are potentially profoundly fruitful for science itself.

I now indicate twelve ways in which physics is transformed, and improved, as a result of rejecting standard empiricism and proceeding explicitly, and not just covertly, in accordance with aim-oriented empiricism.

1   Instead of there being just two (interacting) domains of scientific knowledge, there are three: (a) experimental results; (b) testable laws and theories; (c) metaphysical blueprints (at any stage the best idea as to how the universe is comprehensible). Instead of being separate from science, metaphysics and philos-

ophy become a vital, integral part of scientific knowledge, essential to the intellectual rigour of science. Physics is transformed into natural philosophy.

2   Instead of having a fixed aim and fixed methods, physics has *evolving aims and methods*. This is essential to scientific rationality, and scientific progress. Even though we know (conjecturally) that the universe is comprehensible, we do not know in what precise way it is comprehensible. This we must find out, by discovering which conjectures about comprehensibility lead to the most rapid growth of knowledge when judged in terms of common, implicit criteria as to what constitutes knowledge. A change in aim, in metaphysical blueprint, can lead rationally to a dramatic change in methods. Given any version of personalism, whether multi-theistic or monotheistic, it is rational to attempt to know and influence the natural world by means of methods that are appropriately applied to powerful people – parents, chiefs, kings. Thus prayer, sacrifice and ritual constitute entirely rational methods. Rejection of personalism and adoption instead of some version of physicalism leads, rationally, to a dramatic change in methods. Furthermore, a change in the version of physicalism that is adopted is almost bound to lead to a change in methods. The relatively recent explosive growth of scientific knowledge and understanding during the last 200 years, or even during the last few decades, has only been possible, I suggest, because of the progressive improvement of the aims and methods of science: the adoption of a good aim for science by Kepler and Galileo, suggesting good methods and their subsequent development. As knowledge has improved, so too knowledge about how to improve knowledge has improved.

3   Standard and aim-oriented empiricism lead to quite different views as to how science ought ideally to develop (see figures 9a and 9b). Our starting point is a tribe accepting without question some version of personalism – a(i) and b(i). According to standard empiricism, the ideally rational way to proceed is to divide knowledge up into the observational and theoretical (a(ii)), discard the theoretical, apart perhaps for heuristic purposes to suggest possible testable theories (a(iii)), and then to put forward and test testable theories (a(iv)), thus creating science. According to aim-oriented empiricism, the first major step is to develop a rival cosmological theory (b(ii)), or to create a society in which two rival cosmologies (or religions) coexist, perhaps as a result of two hitherto distinct tribes living together. Only when this has

been achieved is there even the possibility of making a distinction between theory and observation, theory and common-sense (the latter being what is common to the two conflicting theories). The next major step is to discover that individuals can freely invent new cosmological theories and freely modify existing theories (b(iii)), a step first taken in history, it may be held, by the Presocratic philosophers – Thales, Anaximander, Anaximenes, Xenophanes, Heraclitus, Parmenides, Empedocles, Anaxagoras and Democritus. It is perhaps not surprising that this step should be associated with another, the discovery of the impersonal character of the world, the abandonment of personalism and the slow, laborious development of physicalism (the Presocratics being concerned above all with the central problems of physicalism, namely 'what does not change?' 'what does change?' 'how are the two interrelated?'). For, in a personalistic world, denial of the existence of gods, or of God, might be a dangerous matter, whereas in a physicalistic world, mere denial of the existence of atoms, for example, is harmless. This major step creates a major new problem. How can it be discovered which of a multitude of rival religions and cosmologies is true, especially as all are compatible with common-sense (almost by definition)? The solution to this problem is to stipulate that all acceptable cosmologies must postulate that the world is in some way comprehensible, in the sense explicated above (step b(iv)). All rival cosmologies agree that in any bit of the world there are clues or guidelines as to the character of the whole, and that implicit in imperfect knowledge of a bit of the world there are clues or guidelines as to the character of the whole. Each cosmology attributes a different kind of comprehensibility to the world, and thus holds that a different kind of *methodology* needs to be adopted in order successfully to extend and improve (common-sense) knowledge. The primary way to choose between rival cosmologies is to discover which associated methodology is best able to promote the growth of knowledge, when judged by means of implicit, common-sense terms. The discovery that there is one cosmology and associated methodology that has a far greater capacity to promote the growth of (common-sense) knowledge than any of its rivals is the discovery of science (step b(v)). This discovery was made in the seventeenth century, above all by Kepler and Galileo. The cosmology is physicalism: that which is invariant is postulated to be such that it is capable of being characterized by means of some physically interpreted mathematics, from which (together with initial conditions), descriptions of that which varies can (in principle) be deduced. As Galileo put it, 'The book of Nature is written in the language of mathematics.' The method is to put forward mathematically precise conjectures as to what is invariant through change, for example motion, and to test these conjectures by means of precise observation, experiment, measurement. Kepler's laws of planetary motion, Galileo's laws of terrestrial projectile motion, and Newton's unifying laws of motion and gravitation, are the great early successes of this cosmology and associated methodology. Subsequent science continues to develop and select much more restrictive cosmologies and associated methodologies in much the same way, in terms of their capacity to promote the growth of knowledge, within the much more restrictive physicalist conception of comprehensibility. In this way the aims and methods of science improve as scientific knowledge improves.

Unlike the steps advocated by aim-oriented empiricism, those advocated by standard empiricism are either impossible or disastrous to take. Thus step a(ii) cannot be done, as there can be no basis for distinguishing between theory and observation unless rival theories exist. Step a(iii) is an intellectual disaster, as it involves abandoning all theoretical knowledge, together with the indispensable conjecture that the world is comprehensible. It is not in the least surprising that, after these intellectual disasters, step a(iv) fails entirely to provide a basis for the rational growth of knowledge (as we have seen above). The traditional problems of the philosophy of science that arise if some version of standard empiricism is taken for granted are insoluble because these problems presuppose a context for science – step a(iv) – which is itself the outcome of prior, usually unacknowledged, but absolutely disastrous intellectual developments (a(ii) and a(iii)).

4   According to standard empiricism, the critical study of the aims and methods of science – the philosophy of science – is to be sharply distinguished from science itself, just because ideas as to what the aims and methods of science ought to be are not, in any straightforward sense, empirically testable theories. According to aim-oriented empiricism, if science is to be rational, it is vital that the attempt to improve the aims and methods of science (the philosophy of science) be an integral part of science itself. In dissociating the study of aims and methods from science itself, and pursuing it as a metadiscipline, standard empiricism helps to undermine the very thing it seeks to understand: namely the rationality of science.

Natural philosophy (the discipline that emerges, or is recovered, as a result of integrating science and the philosophy of science in the way required by aim-oriented empiricism), can be represented as being made up of three interacting components. There is *knowledge* K (the sum total of accepted observational and experimental results, testable laws and theories); there is the best conjecture as to how the universe is *comprehensible*, C; and there are the currently accepted best *methods* of scientific inquiry, M. The fixed, meta-rules of natural philosophy specify how these three components ought to interact (see figure 10a): they specify

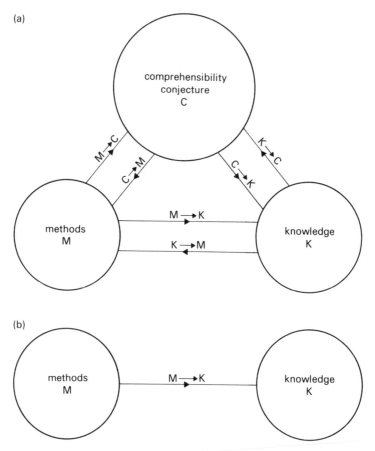

Figure 10(a) Evolving aims, methods and knowledge interacting in accordance with fixed meta-methods of aim-oriented empiricism
(b) Fixed methods of standard empiricism

how we should seek to modify each component to bring it into better accord with the others in such a way as to improve our knowledge and understanding of the world. These six meta-rules (to be established by critical practice and tradition in science) can be indicated as follows. K → M: modify explicit methods M so that they more accurately capture methods which have actually been responsible for the growth of K in history. M → K: modify K by putting M into practice in order to generate new knowledge, and assess existing knowledge. K → C: modify C so that C constitutes the best conjecture as to how the universe is ultimately comprehensible, in the light of all fundamental physical theories, their best interpretation, their common invariance principles, concepts and presuppositions, their inconsistencies and inadequacies. C → K: reformulate the fundamental theories of K in the light of C, so as to accord better with C. C → M: modify M so that M become the best methods to adopt in order to improve knowledge in a universe that is comprehensible in a C-type way. M → C: modify C so that according to C, the universe is comprehensible in just that way which make M the best methods to adopt in order to improve knowledge. The outcome of putting these six rules into practice will be to develop gradually alternative M's, C's and interpretations of K, until eventually alternative testable physical theories are developed. Crucial experiments are then to be performed (as all M's will insist). In comparison with this rich interplay of knowledge, methods and ideals of comprehensibility, the starkness of standard empiricism is glaringly apparent, with its one fixed set of methods M assessing knowledge, and excluding ideas of comprehensibility from the intellectual domain of science on the grounds that such ideas constitute unscientific, untestable metaphysics (see figure 10b).

5   The six meta-rules of aim-oriented empiricism just indicated provide natural philosophy with a rational, even if fallible and non-mechanical, method of discovery. No such thing is possible within the framework of standard empiricism, just because ideas of comprehensibility C are excluded from the intellectual domain of science. It has only been possible for science to make progress because some scientists have put the rational methods of discovery of aim-oriented empiricism into practice despite the prohibitions of standard empiricism. Widespread adoption of standard empiricism has had the consequence that the above six meta-rules of discovery have not been put into practice explicitly and cooperatively by the community of scientists with full understanding of

their rationale. Instead only a few individual scientists have been able to exploit these rules in their work with success: the great theoretical innovators of science, and above all Einstein.

6   Most current versions of standard empiricism – such as those expounded by Popper, Kuhn, and Lakatos – recognize abrupt discontinuities in the development of science at the highest theoretical level. Aim-oriented empiricism, by contrast, recognizes continuity at the highest theoretical level of science. From Thales, Anaximander and Democritus, via the work of Kepler, Galileo, Newton, Dalton, Fresnel, Faraday, Maxwell, Darwin, Boltzman and Planck to Einstein, Schrödinger, Watson, Crick, Salem, Weinberg and Gell-Mann, there is the gradual clarification and development of one basic idea, physicalism: there is that which does not change or vary, X, which determines (deterministically or probabilistically) how that which changes, Y, does change. All major theoretical developments of science can be interpreted as enabling us to understand, in ever greater detail and with ever greater precision, how more and more apparently diverse phenomena are the outcome of relatively few different sorts of entities interacting by means of ever fewer different sorts of invariant forces (at present described by the three or four fundamental dynamical theories of modern physics). Even Darwin can be interpreted as solving a major problem confronting physicalism namely: how is it possible for well adapted, purposive life to develop in a purposeless universe?

It should be noted that in order to make sense of the basic idea of physicalism, (that that which does not change *determines* change) we need to be able to interpret appropriate physical theories as attributing necessitating properties or powers to the physical entities they postulate. This in turn requires that Hume's famous analysis of causation, his denial of the possibility of necessary connections between successive events, be rejected. Elsewhere I have shown how this can be done.[2]

7   On the face of it, however, there are abrupt discontinuities between successive more specific metaphysical blueprints of physics – for example, between the corpuscular blueprint of the seventeenth century, the point-particle blueprint of Newton and Boscovich, the particle field blueprint of Faraday, Maxwell and Lorentz, the unified field blueprint of Einstein.

[2]See Maxwell (1968a, pp. 1–25) reprinted in Swinburne (1974, pp. 149–74).

Each of these blueprints can, however, be interpreted as *generalizing* its predecessor (somewhat as Riemannian goemetry generalizes Euclidean geometry), as long as they are all interpreted as specific versions of physicalism. Thus the corpuscle idea that there is an infinitely repulsive force located on the closed, rigid *surface* of each corpuscle is generalized by Boscovich into the idea that there is an alternatively repulsive and attractice force which varies in a fixed way throughout a *volume* about a central point-particle. This blueprint requires that changes be transmitted instantaneously from point-particle to point-particle through space. More generally, the velocity of such transmission may be *finite*, which means the state of the force-field around each particle will vary (depending on the past motions of the particle). In order to take this case into account, we may coalesce all the diverse force-fields of distinct particles together to form *one* force-field, created by, and acting on, point-particles. This point-particle/field blueprint, associated with Faraday, Maxwell and Lorentz, may in turn be modified by eliminating the particles and insisting that the field *interacts* with itself, small, intense regions of the field standing in for point-particles. This, in essence, is Einstein's unified field blueprint.

The single theme of physicalism runs throughout this dramatic *evolution* of ideas: the basic physical entities (or entity), have invariant physical properties, specified by theory, which determine how the entities interact with one another, and evolve in space and time.

One harmful consequence of the standard empiricist prohibition on critical scientific discussion of such basic metaphysical ideas is that physicists have tended to hold onto outdated metaphysical ideas dogmatically, the best current theories being judged to be incomprehensible as a result (incomprehensibility even sometimes being held to be an inevitable consequence of theoretical advance – predictive success being the most that can reasonably be expected from physical theory, according to those who hold such metaphysically unenlightened views). Thus many of Newton's contemporaries (such as Huygens and Leibniz) condemned his law of gravitation an incomprehensible for failing to be a corpuscular, action-by-contact theory. In a sense even Newton took this view. Judged in terms of Boscovich's more general point-particle blueprint, however, Newton's theory is entirely comprehensible. Again, many nineteenth-century physicists sought to interpret the electromagnetic field theory of Faraday and Maxwell in terms of an underlying material substratum or aether, hoping thereby to make the theory comprehensible, appealing in effect to corpuscular or point-particle ideas of comprehensibility. In terms of Faraday's more general field blueprint, however, Maxwell's electromagnetic theory is comprehensible as it stands: the aether, if anything, can only undermine comprehensibility. An analogous situation has arisen in connection with quantum theory, which is often deemed to be inevitably incomprehensible, no more than an algorithm for predicting experimental results, insofar as it cannot be understood in terms of outdated particle or field ideas.

8   It is above all in connection with Einstein's development of special and general relativity that the rational method of discovery of aim-oriented empiricism is first self-consciously put into practice in science with brilliant success. What we find in Einstein's work is precisely the subtle interplay between evolving theory, evolving methods and evolving metaphysical ideas of comprehensibility (or unity), stipulated in (4) above. As a result in part of Planck's quantum theory of blackbody radiation, Einstein became increasingly aware of a fundamental theoretical and blueprint inconsistency in classical physics: the problem of how continuous electromagnetic radiation can interact with discrete, particle-like matter – the problem of reconciling, as it were, Newton and Maxwell, Boscovich and Faraday. Einstein, meta-physically enlightened, accepted the aetherless, field blueprint interpretation of Maxwell's electromagentic theory. He sought to clarify the nature of the conflict between Maxwell and Newton by extracting from each theory a basic principle, the two principles being mutually incompatible. From Newtonian theory he extracted the (restricted) *principle* of relativity: all laws have the same form with respect to inertial reference frames. This can be interpreted as a *methodological* principle, corresponding to the *metaphysical* principle that physical space is such that a material body cannot have a velocity relative to space itself but only to another material body (the decision as to which is in motion, which at rest, being purely terminological and therefore irrelevant from the standpoint of the character of physical laws). This principle of relativity is only sensible to accept if the aether does not exist (since it is reasonable to suppose that rapid motion with respect to the aether would have physical effects, and would thus be detectable). From Maxwell's theory Einstein extracted the physical law: the velocity of light is a constant (it being, as we have seen, basic to the field-theoretical idea that there is some finite velocity of transmission of influences). Einstein's problem had thus become the problem of

discovering how these two postulates can be made mutually compatible. Einstein discovered that the two postulates can be made compatible if Newton's ideas about space and time are appropriately modified. As a result of making measurement of length, time and mass frame-dependent, the velocity of light can be arranged to be the same in *all* reference frames. This is Einstein's special theory of relativity. This theory leads to a new *methodological* principle, Lorentz invariance, which Newton's theory of gravitation fails to satisfy. A relativistic theory of gravitation is needed. The restricted principle of relativity can be generalized to assert: the laws of nature have the same form with respect to all mutually accelerating reference frames. We can make sense of this idea if we postulate that the effects of acceleration, and gravitation at rest, are equivalent (so that an accelerating reference frame is transformed into a stationary reference frame *in a gravitational field*). We thus arrive at the *methodological* and *metaphysical* principle of equivalence of acceleration and gravitation (implicit in a limited form in Newonian theory). Consider now a large, flat, rapidly rotating disc. According to special relativity, a rigid rod transported from the centre to the circumference will seem to shrink. The geometry of the disc, as measured by the rod, will thus be non-Euclidean. Acceleration affects geometry: hence, by the principle of equivalence, gravitation affects geometry. Mass – or, more generally, energy – may be postulated to affect the geometry of space-time. The field equations of general relativity are the simplest equations which encapsulate this idea, specifying precisely how energy modifies the geometry of space-time. Gravitation *is* the energy-induced curvature of space-time.[3]

After developing general relativity, Einstein abandoned his aim-oriented methodology of discovery in one respect: his (correct) conviction that orthodox quantum theory is unsatisfactory led him to ignore it from a heuristic standpoint, whereas he ought to have tried to extract basic, mutually incompatible principles from general relativity and quantum theory, in an attempt to develop a new unified general relativistic quantum theory.

9 Not only did Einstein invent and brilliantly exploit aim-oriented empiricism in his extraordinarily successful scientific work: he also advocated the view. Again and again in his writings

---

[3]For more detailed discussions of Einstein's scientific journey towards special and general relativity see Schilpp (1970); Holton (1973); Pais (1982).

he emphasized the importance of the conjecture for science that the universe is comprehensible, and the profoundly problematic character of this conjecture. Typical are the following remarks. 'The most incomprehensible thing about the universe is that it is comprehensible' (Hoffmann, 1972, p. 18). 'It is the very essence of our striving for understanding that, on the one hand, it attempts to encompass the great and complex variety of man's experience, and that on the other, it looks for simplicity and economy in the basic assumptions. The belief that these two objectives can exist side by side is, in view of the primitive state of our scientific knowledge, a matter of faith. Without such faith I could not have a strong and unshakable conviction about the independent value of knowledge' (Einstein, 1973, p. 357).

10   Einstein's aim-oriented empiricist way of doing physics exercises a profound influence over modern theoretical physics. The heuristic use of existing theories to develop new theories (special relativity and quantum theory being used to develop quantum field theory, quantum chromodynamics); the formulation of fallible symmetry and invariance principles, such as parity (rejected) and gauge invariance; the drive towards theoretical unity so striking in recent developments: all of this is thoroughly Einsteinian in character and, as Wigner has remarked (1970, p. 15), stems from Einstein's own work. Most contemporary physicists have not, however, renounced standard empiricism: as a result there is still no official place within physics to articulate and criticize metaphysical ideas of comprehensibility. The comprehensibility of the universe is taken to be an article of faith, not a central part of the content of current scientific knowledge. One consequence of this allegiance to standard empiricism is the blindness of most physicists to the inadequacies of orthodox quantum theory.

11   Orthodox quantum theory, lacking any consistent idea as to what sort of entity an electron or proton is in itself, is obliged to be a theory about the outcome of *performing measurements* on these entities. This means, as Bohr always insisted, that orthodox quantum theory is made up of two parts, quantum postulates and some part of classical physics to describe the measurement process. Orthodox quantum theory is thus a severely *ad hoc* theory, being made up of two conceptually unharmonious parts. It is an *aberrant* theory, in that it postulates that something peculiar (and non 'quantal') occurs during the process of measurement. It

lacks precision, just because the key notion of 'measurement' is somewhat vague. It lacks explanatory power, in that it must presuppose some part of classical physics, which thus cannot be explained, and it abandons physicalism.

All these defects can be overcome by developing a full micro-realistic, probabilistic version of quantum theory which attributes *propensities* to electrons, protons, etc., a propensity being a new kind of physical property which determines probabilistically how one entity, such as an electron, interacts with another, for example a positron. This notion of propensity generalizes classical, deterministic ideas of physical property and physical entity.

In order to implement this idea, it is essential to specify precisely the micro-realistic, quantum theoretic conditions for something probabilistic to occur (no reference being made to measurement). My suggested solution to this problem can be put like this. Consider a neutron decaying into a proton, electron and neutrino. Orthodox quantum theory predicts that, in the absence of measurement, the neutron persists in a superposition of the decayed and undecayed states – a state of indecision, as it were, as to what has actually happened. I suggest that after a time $\triangle t = h/\triangle E$ the neutron abruptly jumps probabilistically into one or other state even in the absence of measurement, h being Planck's constant, $\triangle E$ being the difference in rest energy between the neutron on the one hand, and the proton, electron and neutrino on the other hand. And more generally, whenever a composite quantum system evolves into a superposition of two or more interaction channels, each channel containing particles of different rest masses, the superposition persists only for time $\triangle t = h/\triangle E$. Measurement is just a special case of this.

This fully micro-realistic version of quantum theory has in principle all the predictive success of the orthodox theory, and none of the disadvantages. The theory is fully in accordance with physicalism (but not determinism). Furthermore, the theory is capable of being distinguished *experimentally* from orthodox quantum theory, certainly in principle, and possibly in practice (Maxwell, 1976a, 1982). Thus, for example, orthodox quantum theory predicts that decaying systems (such as neutrons) decay at a rate that differs slightly from an exponential rate for short and long times (Fonda *et al*, 1978), whereas the micro-realistic propensity version of quantum theory predicts less deviation from the exponential for short times, and none for long times (due to the prediction of probabilistic jumps predicted not to occur by the orthodox theory).

This simple solution to the problem of developing a fully micro-realistic, propensity version of quantum theory has been over-looked for historical reasons. After the advent of quantum theory, during the years 1925–1937, the physics community was polarized into two opposing camps concerning the status and interpretation of the theory. On the one hand, Bohr, Heisenberg and others held that micro-realistic determinism must be abandoned; on the other hand, Einstein, Schrödinger and others held that micro-realistic determinism must be retained (orthodox quantum theory thus being seriously incomplete). Both camps took for granted that *micro-realism* and *determinism* stand or fall together. As a result, no one at the time had the idea of *defending* one, and *rejecting* the other. No one sought to develop, or even put forward as a possibility, the view advocated here – *micro-realistic probabilism*. This in turn led everyone to overlook the *problem* that confronts micro-realistic probabilism – the problem, that is, of specifying precise micro-realistic, quantum-mechanical conditions for prob-abilistic events to occur. Thus no one put forward the simple *solution* to this problem indicated here.[4]

According to this micro-realistic propensity viewpoint, the mysteriousness of quantum objects, such as electrons and protons, is due to the fact that the basic physical properties of these objects – namely propensities – are of a kind quite different from the basic physical properties of objects we are familiar with from our experience of the macroscopic world, and from classical physics. Familiar macroscopic propensities are not *basic* physical proper-ties. Thus the propensity of a die to be unbiased when tossed, can be explained away in terms of basic deterministic properties of the die and the environment, and probabilistically varying initial conditions from toss to toss. Quantum objects – or *smearons* as I have called them elsewhere – seem strange to us for reasons that are similar to (and no better than) the reasons which led Newton and others to find the Boscovich point-particle mysterious, and Maxwell and others to find the Faraday field mysterious. An electron is a smeared out wave-packet: what is smeared out, however, is the propensity of the electron to interact in a (probabilistic) particle-like way, should physical, micro-realistic,

---

[4]It is above all Popper who subsequently stressed the possibility of developing a realist, propensity interpretation of quantum theory: see his 'Quantum mechanics without the "Observer" ' in Bunge (1967, pp. 7–44). Popper's propensity interpretation of quantum theory fails however to be micro-realistic, in that for Popper *'Propensities are properties of neither particles nor photons nor electrons nor pennies. They are properties of the repeatable experimental arrangement'* (p. 38).

quantum-mechanical conditions to do so arise. All this would have been obvious to the physical community long ago had it put aim-oriented empiricism into practice. Micro-realistic probabilism is at present unknown to the physics community because of its general adoption of standard empiricism, with its exclusion from physics of sustained imaginative and critical metaphysical thought about problems having to do with the overall *comprehensibility* of the universe (of the kind indicated here).

12   From a general intellectual and cultural standpoint, the most important contribution that science has to make is to our overall understanding of the world and our place in it. It is this that is emphasized as being of supreme intellectual importance by aim-oriented empiricism: the whole *raison d'être* for natural philosophy, set within the context of the philosophy of wisdom, is to help people and societies to acquire such understanding as an integral part of life. Standard empiricism devalues, and even suppresses, such understanding for being philosophical, metaphysical, unscientific. As a result, standard empiricism does nothing to heal, and much to promote, the gulf that has grown up between esoteric, expert scientific knowledge on the one hand, unscientific personal understanding of the world on the other hand.

From the standpoint of the overall argument of this book, two implications of the present chapter are especially important. The first is that we must take seriously the thesis that the universe is perfectly comprehensible, as conceived of by physicalism (that is, more or less as conceived of by modern theoretical physics). For this is not a mere metaphysical or philosophical speculation: it is a secure part of scientific knowledge – a presupposition of all of physical science and thus more secure than any particular physical theory. But if the universe is perfectly comprehensible in this way, severe constraints are placed on what can be of value in existence and how what may be of value can be realized – so severe that it may be doubted that anything of value can exist at all (in that all of human life must comply with some unified pattern of physical law). But secondly, if the universe really is comprehensible more or less as modern physics conceives it to be, then one human endeavour, of great value, has been extraordinarily successful: the endeavour of improving scientific knowledge and understanding of the universe. This extraordinarily cooperative, progressive success has been achieved in part as a result of putting into practice aim-oriented rationalism (in the form of aim-oriented empiricism), even if this has not always been recognized. There is here an enormously important methodological lesson to be learned for all human endeavours – a point to be developed further in the next chapter. Just that which seems to pose the greatest threat to the very possibility of human freedom and value – the immense success of physics – actually holds a vital clue for the growth of human freedom and value in all that we do.

CHAPTER TEN

How Can There Be Life of Value in the Physical Universe?

The argument of the last chapter serves to intensify an already severe problem. If the world really is more or less as modern theoretical physics conceives it to be, if it really is comprehensible in the way that physics presupposes it to be, so that some version of physicalism is true – then how can human life have any real meaning or value? If we are all merely very complicated physical systems, made up of molecules in turn made up of electrons, protons and neutrons (electrons and quarks) which interact in accordance with a fixed, unified pattern of physical law, how can we also be people, sentient and conscious? How can it be that we feel, think, enjoy and suffer? How can we be responsible, in any degree for our actions? How can our lives be imbued with any meaning or value? What becomes of our minds and our souls? What becomes of the entire world of human experience if the universe really is more or less as modern theoretical physics seems to tell us it is? The aim to improve our understanding of the world, pursued in order to enrich life, seems to threaten to annihilate conceptually all meaning and value in life. How is this problem to be resolved?

This chapter is devoted to this problem. I begin with some remarks about what is of value in life. I then endeavour to show how that which is of value, which I associate with the experiential realm, may be accommodated within the physicalistic view of the world.

I now endeavour to characterize, in the form of a theory, a *conjecture*, some general features of what is of value in existence. I have fourteen points to make.

1 All that which is of value in existence has to do with life and

above all, for us, with our lives, with human life. We participate in that which is of value in our living, experiencing, doing. Insofar as non-living things are of value, whether natural or made by people, their value arises as a result of their association with life.

2   That which is of value in existence, associated with human life, is inconceivably, unimaginably, richly diverse in character. That which is of most value in one person's life is inherent in the rich pattern of particularities of the person's life, the extraordinarily intricate pattern of environment, deeds, perceptions, feelings, thoughts, desires, imaginings, relationships with others. The greatest poets, novelists and dramatists – Shakespeare, Chekhov, Tolstoy, Stendhal, Jane Austen, Dickens, D.H. Lawrence – can only hint at the rich diversity of value inherent in a person's life. In order to come to see and to understand something of what is of value in another person's life we need the empathetic, imaginative and creative resources of a great artist so that we may enter into the person's world and, in imagination, see, feel, experience, desire, fear, love and suffer what he or she does. We need to acquire deep person-to-person understanding of the other. We need to be an intimate friend at least. It is for this reason that each one of us can glimpse only a minute fraction of all that there is of value in the world in human life.

The richness and diversity of value is made possible, from a neurological standpoint, by the vast structured complexity of our brains, with their 10^{10} neurons. A casual perception, a fleeting thought or feeling, of any person in life has a beauty and profundity greater by far than that of even the greatest works of art, such as a tragedy by Shakespeare, a mass by Bach, a symphony by Mozart or Beethoven. What we are is greater by far than what can be expressed by even the greatest artistic skill. As we live we may deepen our awareness and appreciation of this wealth we inherit, in ourselves and in others: or we may become progressively deaf and blind to it.

In acknowledging the inconceivably rich diversity of what is of value in the world I am not acknowledging that peoples' diverse beliefs about what is or value are all as good as each other. I am not, in other words, putting forward a relativistic or subjectivist view about value. Quite to the contrary, the view that I am proposing is an objectivist, realist view of value. I seek to specify a few general characteristics of that which actually exists in the world, whatever anyone may believe or not believe, that is of value. The value-judgements of some do better justice to what really is of value in existence than the value-judgements of others. In endeavouring to assess the adequacy of any value-judgements, or systems of such judgements, whoever or whatever may uphold them, we need to try to assess how adequately the judgements correspond to the value-reality they are about. Fundamentally, it is not a question of assessing them in terms of some other set of value-judgements, upheld by some other person or being whether it be the majority, the government, the masses, society, culture, the church, history, posterity, the Bible, a prophet, or God.

It is most important to distinguish between diversity in value-judgements reflecting the diversity in what actually exists that is of value, and diversity in value-judgements reflecting disagreement as to what is of value. Some may adopt value-relativism because of a failure to make this distinction, rejection of any system of value-judgements being interpreted, as a result, to be an undesirable refusal to recognize the rich diversity of value.

It should be noted that if what has been said above about the rich diversity of value is true, then it follows that all publicly communicable theories about the general character of what is of value in existence – such as the theory formulated here – must inevitably fail to do justice to what is of value in all its rich, diverse, particular actuality, as experienced by each one of us in our lives.

3   My reasons for rejecting value-subjectivism and value-relativism are extremely simple. To say that any set of value-judgements is as good as any other is, I suggest, equivalent to saying that in reality nothing is of value. For if in reality some things are of value, those value-judgements which recognize this must be better than those which do not; different value-judgements cannot all be equally good, and value-subjectivism and relativism must be false. For value-relativism and subjectivism to be true, in other words, it is necessary for nothing in reality to be of value.

Value-subjectivism and value-relativism also of course annihilate the possiblity of there being genuine individual and cooperative *learning* concerning what is of value in existence. Inquiry pursued in accordance with the philosophy of wisdom, devoting research to learning about what is of value – what it is and how it is to be realized – becomes a nonsense. There can be a change in one's values, but no real learning about what is of value.

It deserves to be noted that there is a close association between value-subjectivism and Cartesian dualism. For of course Cartesian dualism implies that the objective, material world is denuded of value features, just as it is denuded of sensory features. From the standpoint of Cartesian dualism, our experiences of sensory and value features of things in the world around us are all hallucintions, since these features do not exist in reality, in the material world.

4   Our ability to achieve that which is of value must inevitably be limited. Some suffering, failure, injustice is intrinsic to life and cannot be avoided. However fortunate and wise we may be, inevitably in our life we will encounter limitations, failure, misfortune. And there will always be those less fortunate and wise than ourselves. The tragic dimension to life is permanent and unavoidable.

5   Our ability to achieve that which is of value is also limited in a more desirable way. Much that is of value has come into existence unforeseen and unintended. We cannot hold ourselves to be exclusively responsible for all that is of value. Indeed, even when we consciously create something of value, we do so only insofar as Nature, that which is not us, conspires with us to bring about what we intend. The creation and development of human life – the supreme source of value – is almost entirely out of our hands. Our continuing existence, our simplest deeds and thoughts, require the cooperation of Nature in a multitude of ways of which we are ordinarily quite unaware, and even do not understand (in that we are unaware of, and do not understand, the workings of our brains). There may even be a sense in which Nature is wholly responsible for all that is of value in existence; for if, as I shall argue below, the whole human world is a part of Nature, then ultimately it is Nature which produces all that is of value.

6   The ability to experience, participate in and help create that which is of value does not arise abruptly, inexplicably, out of nothing; rather it gradually evolves in time. Sudden blossomings of value owe their existence to long periods of prior germination and growth. We owe our present ability to participate in that which is of value to the actions and efforts of millions of people who have gone before us. Almost everything of value is inherited from the past. Creation is the modification of what already exists. Our present ability to speak, to think, to be conscious and self-aware – our humanity, our self identity as persons – is, as it were, acquired from others: these things develop for us because they have already developed for others. Our existence today depends on a long

process of past social and cultural evolution – and on a long process of natural evolution as a result of random variation and natural selection during millions upon millions of years. It is above all the consideration that we are a part of Nature which compels us to recognize that what is of value evolves gradually in time: abrupt creation of value out of nothing would be inexplicable, a miracle, a violation of natural law.

In seeking to discover and achieve what is of value, our task then is to develop that which is of value which already exists and has been inherited from the past. All attempts to create what is of value by means of abrupt revolutions or conversions which wholly repudiate the past are doomed to failure.

At first sight unprecedented, revolutionary achievements in the arts and sciences – achievements such as those of Shakespeare, Mozart, Beethoven, Newton or Einstein – may seem to tell against the point just made. Closer examination reveals that this is not the case. Shakespeare's plays required the prior existence of Elizabethan society, culture and language, an already developing tradition of poetry and theatre. And most of Shakespeare's plays are based on traditional or historical themes, and modify pre-existing literature. Mozart and Beethoven both required for their work pre-existing musical traditions. Newton himself correctly declared: 'If I have seen further than others, it is because I have stood on the shoulders of giants'. Newton achieved a grand synthesis of the work of Kepler, Galileo, Descartes and many others. Einstein's great contributions to science not only presuppose the whole framework of classical physics, the product of cooperative labour of many people over centuries but also his contributions owe much of their importance to the fact that they resolve problems buried deep in traditional classical physics and mathematics.

7   In order to detect and help create what is of value in existence we need all our personal resources of intelligence, experience, courage and generosity. Above all we need ourselves to feel and to desire. Devoid of feelings and desires, we can only parrot the value-discoveries of others. No scientific instrument or artefact can, as it were, detect the value aspect of reality. In order to be aware of the existence of what is meaningful and of value in the world, it is necessary to be a person responding emotionally to what exists. (This point was the great discovery of Romanticism.)

8   Just as the rich diversity of that which is of value should not

lead us to abandon value-realism, so too the fact that it is necessary oneself to experience, feel and desire in order to perceive value in the world should not lead us to conclude that value is only subjective, not a part of objective reality. On the contrary, what is of value has to do with objective aspects of reality, of people and things in the world, which we perceive through our personal emotional responses to them. There is an analogy here with the perceptual qualities of things. Colours, sounds, smells, as perceived by us (rather than as described and understood by physics), require for their detection the having of appropriate sensations of colour, sounds, smells; but this does not mean that colours, sounds, smells merely are these sensations. On the contrary, these perceptual qualities are, I maintain, objective properties of things perceived by us via our sensations. Similarly, that which is of value exists objectively in the world; it is perceived by us through our emotional responses to things.

On the one hand, a blade of grass can be objectively green even though no one happens to perceive its greenness. On the other hand, a person may experience visual sensations of greenness and yet not perceive any green thing, in that he experiences merely an after-image, an optical illusion, a hallucination.

Analogously, on the one hand, a piece of music may be objectively beautiful even though no one happens to experience its beauty; a human action may be objectively noble or cruel even though no one happens to experience or perceive the action in this way – possibly not even the person who performs the action. On the other hand, things may seem to be desirable, beautiful, cruel or ugly – they may be experienced in these ways – and yet in reality may not be these things. Just as our sensations, our inner experiences, can lead us to misperceive what is before our eyes, so too our emotional responses, our 'value-perceptions', may delude us about the significance and value of what goes in the world around us. Indeed, value illusions and hallucinations are probably far more common in life than illusions and hallucinations of a kind that would ordinarily be described as having to do with objects and facts.

We have to learn to see aspects of the world around us: stones, people, trees, sky. Equally, we have to learn to see meaning and value in the world around us, in our environment, in events, in human actions and lives. As I have already indicated, inevitably the full richness, significance and value of what there is in the world escapes human perception and understanding. A person dies. Something infinitely precious has ceased to exist. Almost certainly, however, no one is aware of the full significance of the person's life. Even an intimate friend, a lover, can only know of aspects of the value of the person's life. Even the person herself probably failed to appreciate adequately her own value. The full significance and value of the life is something that eludes us all: and yet it is something that did objectively exist in the world, in the realm of actuality.

9   I have already remarked that our feelings and desires, though necessary for the perception of value, are not infallible guides to what is of value. Equally, no prophet, religion, revelation, book, tradition or institution is an infallible guide to what is of value – just as none of these constitute infallible guides to truth, to knowledge. Our attitudes to traditional judgements concerning what is of value ought to be analogous to attitudes to traditional scientific judgements concerning truth encapsulated in our best scientific theories: these traditional judgements, even if the best we have, nevertheless are no more than fallible, imperfect conjectures, always open to development and improvement.

10   In addition – it almost goes without saying – there are no infallible methods or recipes for the achievement or creation of that which is of value. We cannot infallibly achieve value or know we have achieved it – even when the achievement has actually been made.

11   The inevitability of doubt about the meaning and value of our lives ought not to be the cause of despair – any more than the inevitability of doubt in science ought to be the cause of scientific despair. Acknowledging calmly the inevitability of doubt about the meaning and value of our lives makes learning and growth possible, just as in science. Repudiation of doubt, out of fear, obstructs learning and growth. We should not seek to *rebut* scepticism about value: rather we should seek to *exploit* it in an endeavour to help increase value. As in science, so in life: we need to be so unrestrictedly sceptical, in our endeavour to realize what is genuinely of value, that we become sceptical even of the capacity of unlimited scepticism to promote the realization of value. As in science, so in life: total scepticism is to be rejected on pragmatic grounds; it cannot help. We are rationally entitled to assume that our lives here on earth are genuinely meaningful and of value, even though this cannot be verified or proved, just as we are rationally entitled to assume that the universe is, in some way,

comprehensible even though this cannot be verified or proved. These are the only two basic rationally justifiable articles of faith – justifiable on pragmatic grounds, without one iota of scepticism being repudiated. Our problem is to reconcile these two basic tenets of rational, sceptical faith. How can there be life of value in this physically comprehensible universe? This is a *practical* problem (a problem of acting so as to realize value in this world), a *theoretical* problem (a problem of choosing the best answer from many possibilities) and a *conceptual* problem (a problem of discovering just one possible answer).

12   In order for that which is of most value actually and potentially in existence to flourish, we need to endeavour cooperatively to improve our aims and methods as we live – seeking, in this way, to put aim-oriented rationalism into practice in the world. Cooperative aim-oriented rationalism provides a framework within which diverse philosophies of value – diverse religions, political and moral views – may be cooperatively assessed and tested against the experience of personal and social life. There is the possibility of cooperatively and progressively improving such *philosophies of life* (views about what is of value in life and how it is to be achieved) much as *theories* are cooperatively and progressively improved in science. In science diverse universal theories are critically assessed with respect to each other, and with respect to *experience* (observational and experimental results). In a somewhat analogous way, diverse philosophies of life may be critically assessed with respect to each other, and with respect to *experience* – what we do, achieve, fail to achieve, enjoy and suffer – the aim being so to improve philosophies of life (and more specific philosophies of more specific enterprises within life such as government, education or art) that they offer greater help with the realization of value in life. It is of course true that we understand and judge what we do, the extent to which we succeed and fail, even our enjoyment and suffering, in terms of our explicit or implicit philosophies of life. As a result, experience and philosophy may simply reinforce each other to produce dogmatism, and failure to see even the need for learning. An analogous situation arises, however, in connection with science: observations and experiments are interpreted and judged in terms of theory, there thus always being the danger here too that experience and theory uncritically reinforce each other to produce dogmatism. The solution in both cases is to consider a number of rival universal ideas (theories or philosophies) there being tripartite assessment

between idea, idea and experience. For this to occur, in science or in life, sympathetic person-to-person understanding needs to develop between individuals, and between theories and philosophies (or cultures). In this way, multiplicity of religions, philosophies, cultures, ways of life, can be enriching for us all (just as multiplicity of theories can enrich science), instead of such multiplicity being, as at present, a source of incomprehension, fear and conflict.

13 At present our thinking about what is of value, and our real-life capacity to realize what is of value, are seriously obstructed by a wide-spread tendency to run together doctrines that ought to be sharply distinguished. At its most extreme, this tendency may be delineated as follows. There is, to begin with, a tendency, in some religions and political doctrines to amalgamate the following five distinct doctrines: (1a) it is not the individual that is of supreme value, but rather something else (one particular individual, some individuals, God, the state, the masses, humanity, or an ideal society in the future); (2a) the individual ought not to decide for himself what is of value; rather he should allow this to be decided for him by whatever it is that is of supreme value (a religious leader, a group of religious leaders, God, the church, tradition, history, a sacred book, the state, reason, science, the masses, the majority, humanity as a whole, future opinion); (3a) the individual should not value what he desires but on the contrary should recognize that his desires (or most of them) are in opposition to what is of value; he should not selfishly pursue his own interests, but on the contrary should seek to serve what is of supreme value rather than self; (4a) value exists objectively; (5a) what is of value in existence is beyond doubt: it can be known to be of value with absolute certainty.

Any doctrine which amalgamates these five doctrines may be called a version of *authoritarian objectivism*. Anyone who upholds liberalism in the broadest sense – in that they value individual liberty and tolerance, and reject authoritarianism and dogmatism in all its forms – must find authoritarian objectivism abhorrent. Failure to distinguish carefully the five different doctrines that go to make up authoritanian objectivism will however lead the anti-authoritarian liberal into rejecting all of them; as a result, he will come to uphold an opposing doctrine, which may be called *individualistic subjectivism*, which amalgamates the following five opposing doctrines: (1b) it is the individual that is of supreme value; (2b) the individual ought to decide for himself what is of value, as far as he is concerned; (3b) the individual should value what he desires; he should seek to satisfy his own desires; he should selfishly pursue his own interests; (4b) value is subjective; (5b) what is of value for each one of us is a legitimate matter of doubt.

As an attempt at something better than authoritarian objectivism, individualistic subjectivism is a disaster, in that it is inconsistent and self-defeating. Almost all the disastrous defects of individualistic subjectivism come from the rejection of value-objectivism (4a), and the affirmation instead of value-subjectivism (4b).

Thuus, upholding subjectivism (4b) sabotages just that which the tolerant liberal is most concerned to affirm, namely the *objective* supreme value of the individual (1b) – an objectivist conjecture about what is of value in the world. In short, (4b) is incompatible with (1b). If (4b) is true, and all value-systems are equally viable, then so too are those which deny (1b) and assert some form of authoritarian objectivism. The value-subjectivism of individualistic subjectivism annihilates all reasons for repudiating authoritarian objectivism.

It deserves to be noted that (1b) is also incompatible with (3b): for if what is of supreme value is the individual person, then individual persons ought surely to recognize this and act accordngly, by treating others as ends in themselves, as Kant, for example, advocates. They ought not to satisfy their own desires if this involves exploiting others.

Again, in advocating value-subjectivism (4b), the tolerant liberal may hope, in this way, to oppose those doctrines in authoritarian objectivism that are for him the most abhorrent – the authoritarianism and intolerance, the demand for submission, obedience and self-sacrifice, explicit or implicit in (1a), (2a) and (3a). His hope is that if subjectivism comes to be generally accepted, it will lead to general tolerance, to respect for the diversity of values and ways of life, in that people will cease to claim that what they value is better than what others value, that they have a moral duty to conquer and convert others – and will thus cease to strive to conquer and convert in actuality.

The aim here is excellent, but the means chosen to realize the aim are disastrous. Value-subjectivism, far from helping to oppose authoritarianism and intolerance, actually destroys the only rational basis for opposing these things. For, from the standpoint of value-subjectivism (4b), one is obliged to hold that although these things may be immoral with respect to one value-system, they will nevertheless be highly moral with respect to other, equally good, value-systems.

Yet again, in advocating value-subjectivism, the tolerant liberal may hope to promote anti-dogmatism (5b) and oppose dogmatism (5a). But, once again, the thing is all the other way round. Value-objectivism (4a) provides the only rational basis for doubt about what is of value. Subjectivism annihilates the very possibility of doubt: it actually renders doubt meaningless, since subjectivism cannot make sense of the idea that a value judgement is in any sense wrong. Thus for subjectivism, there can be no valid role for reason or for criticism within the realm of value; and there can be no learning. In all these respects, individualistic subjectivism is actually worse than authoritarian objectivism. This latter doctrine can at least hold doubt to be meaningful (even if it deserves to be rejected); it can give a role to reason and criticism, and can acknowledge the possibility of learning. Thus, once again, the rejection of objectivism, and the adoption of subjectivism in its stead serves to annihilate the very values the tolerant liberal seeks to affirm.

Finally, of course, value-subjectivism implies that nothing is objectively of value. The failure to distinguish the five distinct doctrines of authoritarian objectivism has the disastrous consequence that we leave open for ourselves only the following extraordinarily unattractive choice: either we must accept the supreme value of something other than ourselves, to which we must sacrifice our intellectual independence and our individual freedom: or we must accept that everything is ultimately meaningless and valueless. This is not a desirable choice to be forced to make. Insofar as individualistic subjectivism is the official opposition to authoritarian objectivism, one might almost suppose that it had been deliberately cooked up by authoritarians to be as grim as possible to discourage as many people as possible from leaving the prison of authoritarianism.

In developing a better alternative to authoritarian objectivism than individualistic subjectivism, the decisive point to recognize is that value-objectivism, or value-realism, far from being naturally aligned with dogmatism and authoritarianism, is actually incompatible with these things, value-realism actually being essential to provide a rational basis for doubt, and for learning, in the realm of value. Furthermore, the five components of authoritarian objectivism need to be sharply distinguished, and dealt with separately, one by one. The result, I suggest, is the doctrine proposed here, in points (1) to (12) above. It might be called conjectural, cooperative, experiential realism, or experiential realism for short, it being understood that cooperativeness and individual freedom are interdependent as explained in chapter 8.

It deserves to be noted that this discussion of the relative merits and demerits of *authoritarian objectivism, individualistic subjectivism* and *experiential realism* is highly relevant to the overall theme of this book. Following Plato, the academic enterprise is at present, one might say, committed to *authoritarian objectivism*: it is recognized, however, that committing the academic enterprise to this doctrine in the realm of value would be intolerable (because of its authoritarian, anti-liberal, Platonic consequences): hence the academic enterprise restricts itself to presupposing authoritarian objectivism in the realm of fact. This is married to acceptance of *individualistic subjectivism* in the realm of value – the division aided by past acceptance of *Cartesian dualism*.

I propose that both authoritarian objectivism for the realm of fact, and individualistic subjectivism for the realm of value be rejected, and that instead we adopt *conjectural physicalism* as far as the physical universe is concerned, and *conjectural experiential realism* as far as the human world is concerned – the world as lived and experienced by us, imbued with human enjoyment and suffering, human meaning and value.

The above account of some general features of what is of value in the world is put forward as providing a conjectural background framework for more specific and definite conjectural philosophies of value. Do I have any such more specific philosophy of life to offer, compatible with but more definite than, the points (1) to (13) above? In an attempt to answer this question, I put forward the following slightly more definite conjectural philosophy of life.

14   The poles of value are life and love on the one hand, suffering and death on the other hand. The supreme good in existence is living life lovingly, actively loving that which is lovable in existence; and the supreme evils are suffering and death. Everything else of value in existence is organized around these two poles of good and evil.

We can help our love to grow, or to wither and die, by what we do, what we attend to, what we strive for and value. We cannot, however, authentically command ourselves to love X, or decide to love Y at will, since real love is too dependent on spontaneous, instinctive feeling and desire, out of our immediate control. We cannot therefore sensibly demand of ourselves, and of each other, that we should indiscriminately love our fellow human beings. We can however sensibly strive to create a world in which people, on the whole, treat each other, and do things together, in ways which are in accordance with certain necessary conditions for love to exist. Thus we can strive to create justice, democracy, individual freedom, tolerance, cooperative rather than hierarchical social structures, traditions of resolving conflicts based on mutual understanding, good will and cooperation rather than on bargaining, manipulation, threat or violence. In this way, love can be held to be the supreme positive value, from which all others are, as it were, derived. Justice, peace, cooperativeness, democracy, health, prosperity, enjoyment, knowledge and understanding, reason, creativity, skill, imagination, courage, beauty, sensitivity, compassion, cherishing, active concern for one's own welfare and for the welfare of others, generosity, friendliness, freedom, passion, life itself: these are all of value insofar as they are necessary conditions for the supreme thing, love.

But in addition we may hold that suffering and death are evils in their own right, as it were, and not evil only insofar as they negate the possibility of love. We do not need to appeal to the value of love in order to provide a rationale for striving to avoid unnecessary suffering and death: these endeavours carry with them their own rationale. Attempts to cure and prevent disease, to end war, totalitarianism, torture, exploitation, poverty require no further *raison d'être* than that of bringing avoidable suffering and death to an end.

Ideally, then, we live life lovingly, and in such a way as to minimize suffering. The fundamental purpose of academic inquiry is to help us develop a less suffering, more loving world.

With these preliminaries over, we come now to the central problem of this chapter. How can the rich world of human experience, full of colour, sound, love and hate, joy, tedium and pain, imbued with meaning and value – the world of Shakespeare, Tolstoy, Chekhov, Mozart, Bach, the Renaissance, the French Impressionists – be accommodated within the physical universe, as conceived of by modern physics? How can we be conscious, free and loving persons if we are merely electrons, protons, and neutrons interacting in accordance with precise physical law? How is it possible to reconcile *physicalism* and *experiential realism* (as set out in the above fourteen points)?

This is an old problem. It goes back at least to Democritus. An important part of the problem can be put like this. If physicalism is correct, then the world is such that it is in principle possible to formulate a testable unified physical theory, which is both *true* and *complete*. This theory – let us call it T – would in principle apply to all that there is, and would in principle predict and explain all phenomena. (In practice, of course, it would be possible only to apply the theory to the very simplest of phenomena, to only a limited degree of accuracy.) The theory would give a precise specification of the physical nature of the few fundamental physical entities of which everything is composed, and would specify precisely the nature of the invariant property (or properties) possessed by these entities, determining how what changes does change. Conceivably, there might be just one entity with one invariant property determining how what varies does vary, in space and time. T would unify all the forces of nature; it would unify general relativity and current quantum field theories of the electromagnetic, weak and strong forces. T, we are supposing, is true, complete and comprehensive: and yet, it seems, it could not predict the content of the world of human experience – colours, sounds, smells, tastes, tactile qualities of things, as experienced by us; our inner experiences, sensations, feelings, thoughts, states of consciousness; the vast diversity of human character and person-ality; the beauty, tragedy and value of human life; its joys and pains, its inner meaning.

Descartes sought to solve this problem by, in effect, conceding that (1) T gives a true, complete and comprehensive account of what there is in the world of matter; but at the same time postulating that (2) there is a distinct world of mind, which accommodates all that there is in the world of human experience which T fails to predict. For each one of us, our distinct, private 'world of mind' is linked in some way with our brain. These two postulates, (1) and (2), constitute the essence of Cartesian dualism. They do provide some sort of solution to the *physical universe/human world problem* (as we may call our original problem), but only at the expense of creating a number of severe new problems, such as (a) how is the mind related to the brain? (b) how can the mind influence the brain, if T is true? (c) granted that the mind cannot influence the brain, and that it is the brain that controls our bodies, how can there be free will? (d) if all we are ever aware of is our own private world of consciousness, our world of mind, how can we ever come to know anything of the external world of matter?

Since Descartes' time, the main effort of western philosophy has been to solve these and related problems that arise once some form of Cartesian dualism is accepted. Locke, Spinoza, Leibniz, Berkeley, Hume, Kant, Hegel, Schopenhauer, Mill, Mach, G.E. Moore, Russell, Wittgenstein, Schlick, Feigl, Ayer, Ryle, Popper, Smart, Armstrong, Fodor and many others, have been centrally preoccupied with aspects of the problems generated by Cartesian dualism even when, as has often been the case, dualism itself has been repudiated. One striking feature of this tradition of discussion is a tendency to lose sight of the nature of the original problem – the physical universe/human world problem – which Cartesian dualism sought to solve.

Instead of attempting to solve the problems generated by Cartesian dualism, what we need to do, I suggest, is return to the original physical universe/human world problem, recognize clearly that Cartesian dualism fails to solve this original problem, and develop a better resolution of the problem.

I suggest that both (1) and (2) above, of Cartesian dualism, are false. It is false that (1) T gives a true, complete and comprehensive account of what there is in the world of matter. Rather T gives a true, complete and comprehensive account of what there is in the world of a very special, highly restricted kind. T provides us with what might be called a skeleton description of the world. Only properties of a very special kind are described by T – or by descriptions of states of affairs formulated in the vocabulary of T. In particular T leaves out all mention of the experiential qualities of things – their colour, sound, smell, feel, as experienced by us, and their beauty, ugliness, meaning and value as felt and experienced by us. T is specificially designed to omit all references to such qualities of things: thus the fact that T is silent about them provides us with no reason whatsoever for supposing that they do not really, objectively exist, as a part of the material world. *Physicalism and experiential realism can thus both be true.* The world of physics and the world of human experience dovetail together to form one unified material world. The very distinction between 'the physical universe' and 'the world of human experience' is, as it were, an artifact of our understanding rather than something that exists in reality.

Not only is (1) false; (2) is false as well. The basic reason for believing in (2) – for believing in a 'world of mind' distinct from the 'world of matter' – was belief in (1), namely that T is *entirely* complete and comprehensive about the world of matter, as it were, it therefore being necessary to postulate an additional world to accommodate all that T leaves out, namely the experiential.

Thus the moment (1) is rejected – the moment it is recognized that T provides a true, comprehensive description of what there is in the world of *only a very special, restricted kind*, (silence about experiential qualities being no grounds for holding they are not a part of the material world) – the basic reason for believing in (2), in the existence of a distinct 'world of mind', collapses. I conjecture that no such distinct 'world of mind' exists. At a stroke, the Cartesian problems that arise from postulating the distinct 'world of mind' vanish as well.

In what way precisely is the *comprehensive* description of T highly *restricted*? If T is ever formulated, it will have arisen as a result of the endeavour to *predict, explain* and *understand* all phenomena. T is designed specifically to specify the unified pattern running through all phenomena, however diverse, controlling how they occur. Thus T may be said to refer only to those *casually efficacious* properties which everything has in common with everything else.

The precise way in which T is both comprehensive, on the one hand, and highly restricted, on the other hand, can be specified as follows.

Given any bit of the world, isolated from the rest of the world, then for any instant t_0, there exists a true description D_0 of this bit of the world, formulated entirely in the vocabularly of T. D_0 specifies the instantaneous physical states of the basic physical entities that go to make up the bit of the world in question – their relative positions, velocities and so on, their masses, charges, momenta, energy or whatever. Furthermore, the true description D_0 is such that it, together with T, predicts future states of the bit of the world in question at future times t_1, t_2, when described in precisely the same way, i.e. in terms of physicalistic descriptions D_1, D_2. T is comprehensive and complete in the sense that it refers to *all* the invariant physical properties determining change in *all* possible isolated bits of the world. But this does not mean that T, D_0, D_1 or D_2 tell us all that is true of the bit of the world in question. Only those properties will be referred to and described which need to be referred to and described in order that the above predictive tasks become (in principle) possible. Thus, if omission of all reference to experiential qualities does not in any way disrupt the capacity of T plus D_0 to imply D_1, then experiential qualities will not be described by D_0 or D_1. If, for example, the bit of the world in question contains a person, who sees a blue light at time t_0 which turns into a red light at time t_1, then D_0 and D_1 will describe the light and the person as light of such and such wavelengths, and electrons, protons and neutrons (which make up atoms, molecules and neurons of the person's brain) interacting in such and such ways: there will be no reference whatsoever, however, to the experiential qualities of blueness, redness, and no reference to visual sensations of blue and red, just because omitting all reference to these experiential qualities does not disrupt physicalistic prediction. T applies to all that there is, and predicts everything that occurs when described in terms of the highly specialized, highly restricted vocabulary of T: it does not, however, tell us all that is true of what there is. In particular, T tells us nothing whatsoever about the rich, diverse experiential dimensions of reality (of which we only catch a glimpse in our own experiences).

A further decisive difference between physicalistic propeties (of the kind referred to by the vocabulary of T) and experiential qualities, is the following. In order to come to know what sort of property any physicalistic property is, it is not necessary oneself to have any special sort of experience (although some experiences are obviously necessary if one is to be sentient and conscious, and thus capable of understanding anything). In order to understand the meaning of any physicalistic term – such as mass, electric charge or whatever – it is not necessary oneself to have had any special sort of experience. In sharp contrast to this, in order to know what sort of property an experiential propety is, it *is* necessary oneself to have had certain sorts of experiences. In order fully to understand the meaning of experiential terms such as red or blue, or love, despair, kindness etc., it is necessary oneself to have had certain sorts of experiences. Being blind from birth does not in itself debar one from understanding any physics: it does however debar one from understanding the visual part of the experiential domain. Physical colour can be understood; experiential colour cannot.

At this point it may be asked: given that T has been formulated, why cannot one formulate a new, even more complete and comprehensive theory T′, which would consist of the postulates T plus additional postulates correlating physicalistic states of affairs and experiential qualities and states of being? The answer to this is that it might well be possible to formulate such a T′, but a terrible price would be paid. T′ would be so grotesquely complex and *ad hoc* that it would be entirely *non-explanatory*. It would predict, but it would not explain. Just because of the incredibly rich diversity of the experiential world, and because of the incredibly complex way in which this connects up with the physical world (via our 10^{10}

neurons, in different ways for each one of us), one would need to add endlessly many new postulates to form T'. (Endlessly many new postulates would need to be added with the birth and growth of every person.)

We thus have here an *explanation* as to why there cannot be a good explanation as to why physicalistic properties and experiential qualities are correlated in the ways that they are. I suggest that this solves one major part of the problem of reconciling the physical and the experiential. The experiential domain has always seemed profoundly mysterious from the standpoint of physicalism, just because of the apparent impossibility of explaining and understanding the experiential domain. What the above argument does is to explain, and thus demystify, the impossibility of giving good (scientific, physicalistic) explanations of the experiential domain.

Put another way, the above argument shows that in order to explain and understand phenomena in the way that physics enables us to do, by revealing underlying unified patterns in ostensibly diverse phenomena, a certain price must be paid. The rich particularity and diversity of the experiential aspect of reality must be neglected. This can always be put back in again, but only at the price of one's predictive theories becoming so grotesquely complex and *ad hoc* that they cease entirely to be explanatory. Physics is only able to delineate the unified skeleton of the world by leaving out the richly diverse experiential flesh.[1]

An explanation has been given as to why there can be no good explanation of how and why the experiential flesh of the world exists as it does amongst the physicalistic bones – no good explanation, that is, of the kind sought by physics. Might it not be possible, however, to explain and understand correlations between physicalistic and experiential aspects of the world in some other way, in terms of some other notion of 'explain' or 'understand'?

We have already acknowledged the existence of two kinds of understanding – or rather, the existence of two interdependent aspects of understanding. On the one hand there is person-to-person understanding, achieved when one person can imaginatively recreate for himself the view of the world, aims, problems, experiences, desires, hopes and fears of another person, thus entering imaginatively and accurately into that other person's

---

[1]For further aspects of this proposed solution to the physical universe/human world problem see Maxwell (1966, 1968a, 1968b).

experiential world; on the other hand, there is scientific or physicalistic understanding, achieved when a group of people develop an empirically successful theory which attributes a unified pattern to a range of ostensibly diverse phenomena.

I have emphasized that these two kinds of understanding are interdependent. There can be no successful person-to-person understanding without some sort of 'scientific' understanding of the environment in which the person to be understood exists, however primitively pre-scientific this understanding may be. And there can be no scientific or physicalistic understanding of the world without scientists being able to acquire person-to-person understanding of each other in the context of science, to the extent of being able to enter imaginatively into each others' scientific views of the world, research aims and research problems, and scientific experiences (observations and experiments). Scientific theory and knowledge – embodying our scientific understanding of the world – is itself the product of a multitude of past and present person-to-person understandings achieved by scientists of each other.

This interdependence of person-to-person and scientific understandings is not, however, itself generally acknowledged and understood, either by those concerned primarily with person-to-person understanding, or by those concerned primarily with scientific understanding. It never occurred to the 'philosophes' of the Enlightenment to divorce passionate concern for the inner life of man from passionate involvement with the imaginative and critical exploration of the natural world being undertaken by natural science. Romanticism created this divorce. Rousseau, Blake, Wordsworth, Keats, Tolstoy, Kafka, D.H. Lawrence and a multitude of other novelists, poets, dramatists and artists passionately pursued person-to-person understanding – exploration of the experiential world – in a way that was divorced from, if not actually hostile to, science. (Chekhov is a notable exception.) Natural scientists, on the other hand, in conformity with the philosophy of knowledge, developed scientific knowledge as if it were wholly impersonal, something quite distinct from person-to-person understanding. (Here, Einstein is a notable exception.)

The result of these intellectual and cultural developments is to create the impression of two disassociated 'worlds' – the world of physics, the physical universe, on the one hand, and the world of human experience and life, the experiential world, on the other hand. In immersing ourselves in science, we forget, or fail to realize, that the experiential flesh of the world has been deliberately excluded so that unified patterns of law may be discerned in the bare bones of the world: and in immersing ourselves in the experiential world, the world of person-to-person understanding, of history, biography, literature and art, we may find it necessary to repudiate the scientific vision of the world in that it is (mistakenly) interpreted to annihilate the experiential domain.

And there is a further point. Failure to acknowledge the very different (through intellectually equally legitimate and interdependent) physicalistic and person-to-person modes of understanding may itself be a source of mystification concerning the 'comprehensibility' of the experiential. In an important sense, these modes of understanding proceed in opposite directions. What is for one mode the base line of comprehensibility (in terms of which everything else needs to be understood) is for the other mode wholly incomprehensible. Thus for person-to-person understanding, the base line of comprehensibility is made up, for each one of us, of our own personal elemental experiences and actions, in terms of which we seek to 'understand' experiences and actions of others. The physical properties of fundamental physical entities, and the patterns postulated by fundamental physical theory, being utterly remote from our personal experience, are utterly incomprehensible, from a person-to-person standpoint. In terms of physicalistic understanding, however, exactly the reverse of this holds. The physical properties of fundamental physical entities – the unified patterns postulated by physical theory – constitute the base line of physicalistic understanding, in terms of which we seek to explain and understand everything else. From the standpoint of this mode of understanding, it is our personal experiences and actions that are almost inconceivably incomprehensible, being the outcome of interactions of millions upon millions of fundamental particles organized in an incredibly complex and specialized way into the cells of our brain and body – the functioning of the brain, in particular, being profoundly affected in an incredibly intricate way by years of intricate, particular occurrences in the past. If now we fail to distinguish these two opposingly-directed (but interdependent) modes of understanding, and as a result suppose that there is just one uni-directed mode of explanation and understanding, we will inevitably, as a result, be deeply mystified by both the experiential and the physicalistic. Our immediate experiences, in one way so utterly comprehensible, will also seem, in a wholly puzzling way, to be completely incomprehensible. Electrons, photons, protons, quarks, in one way entirely comprehensible, will also seem, in a wholly puzzling way, to be completely incomprehensible.

In order to resolve these puzzlements – a major part, I suggest, of puzzlement concerning how the physical and the experiential are interrelated – we need to recognize clearly the existence of the above two distinct, intellectually equally legitimate, interdependent but opposingly directed modes of explaining and understanding. The view that understanding does have this character may be called the *duo-directional theory of understanding*. This theory must be an important part of our 'understanding' of how the experiential world of human life can be accommodated within the physical universe.

In order to improve further our understanding of how the experiential and physical worlds dovetail together, we need, I suggest, to take the following two steps.

1 We need to recognize that all human life, and indeed all life, is essentially purposive or aim-pursuing in character, person-to-person understanding being a form of purposive understanding, and itself, indeed, exemplifying purposiveness.

2 We need to develop a mode of inquiry that I shall call the *generalized Darwinian research programme*. This seeks to improve our knowledge and our (duo-directional) understanding of how purposiveness has gradually evolved in the world fully in accordance with the (presumed) fixed physicalistic structure of the universe.

I take these two points in turn.

1 We are what we do. All our human world (personal, social, cultural, intellectual, spiritual) is purposiveness, the exemplification of aim-pursuing. Our imagining, thinking, feeling, dreaming is activity, aim-pursuing, as explained in chapter 8. All meaning and value exists only in association with aim-pursuing. Scientific knowledge and understanding themselves constitute aim-pursuing. All this has been argued for throughout this book.

Not all our purposiveness or aim-pursuing is however human life, or indeed life of any kind. A simple feedback device such as a thermostat, or a somewhat more sophisticated feedback device such as a self-guiding rocket, can be regarded as an 'aim-pursuing' device, as the term is used here.

A device 'pursues an aim A' to the extent that almost all possible routes the device might take into the future in the given environment fail to realize the aim A, and yet almost always the device pursues one of those very special routes that do take it to A. A thermostat or self-guiding missile does this by means of feedback mechanisms which adjust the 'direction' of the device's activity in the light of environmental disturbances, so that one of those very rare paths to A is persistently pursued. In the case of a thermostat or self-guiding missile, no problem arises in understanding how aim-pursuing is compatible with physicalism: in these cases successful aim-pursuing actually requires that there is a fixed pattern of physical law which the feedback mechanisms of the device obeys.

Person-to-person understanding is a special case of *purposive* explanation and understanding. The latter explains the actions of an aim-pursuing entity essentially by showing how these actions fit into a pattern of hierarchically organized goal-pursuing. Short-term goals are pursued in a particular sequence in order to realize some overall goal. In this way, actions can be explained as solutions, or attempted solutions, to problems: reasons can be given for actions, and for changes of activity. In explaining and understanding the actions of a self-guiding missile, let us say, in this sort of way, there is however no suggestion that the missile is sentient, let alone conscious: thus person-to-person understanding is not involved. Purposive understanding might be described as a highly etiolated form of person-to-person understanding (the latter being a highly enriched version of the former, applicable to purposive beings that are sentient, conscious, and thus *persons*).

In order to understand how the experiential world of human life, imbued with meaning and value, can be a part of the physical universe, an important step is to acquire (necessarily etiolated) *purposive* understanding of human action and human life. This would include purposive understanding of those goal-pursuing actions that consist of one person acquiring person-to-person understanding of another person. It would also include purposive understanding of those cooperative goal-pursuing actions that consist of scientists improving scientific knowledge and understanding of the world. It would enable us to understand, in purposive terms, aim-pursuing associated with sentience, consciousness, communication and even love, between people: it would enable us to understand how these things can exist and proceed in a way that is in accordance with the physicalistic structure of the universe: and yet it would not itself embody person-to-person understanding. Person-to-person understanding would be, as it were, superimposed on top of purposive understanding of human life.

In this way, it becomes possible to set the sentience, consciousness, meaning, value, love and suffering of human life into a broader, intelligible context of purposiveness – a sort of neutral buffer zone between the experiential world and the physicalistic universe. It becomes possible, in principle at least, to develop purposive understanding of goal-pursuing that very gradually, over millions of years of evolution, becomes sentient, conscious, personal. In terms of physicalistic and person-to-person modes of understanding alone, this cannot be achieved.

2   In order to implement the programme I have just indicated, we need to adopt an historical approach. In particular, we need to improve our knowledge and understanding of the history of human and other life on earth by pursuing a mode of inquiry that I call the generalized Darwinian research programme.

All life is the embodiment of purposiveness. (Plants achieve their goals primarily by growth.) From the present standpoint, Darwin's great achievement was to provide an explanation as to how the vast diversity of forms of embodied purposiveness we find on earth can have come to be even though we live in a physicalistic universe. Darwin postulated two mechanisms: (a) random inheritable variation; (b) natural selection. We conjecture that billions of years ago, molecules developed that acquired the capacity to reproduce: possibly these consisted of crystalline rods which grew in length until they broke as a result of environmental buffeting. Those inheritable variations with the greatest capacity to grow and reproduce multiplied, while other variations died out: this process continued, leading in the course of time to the world as we find it today, including ourselves. Amongst the predictions of the theory are the following. (i) In its given environment, an animal pursues a pattern of goals, a way of life, organized so as to promote the overall goal of reproductive success. (ii) Its body is designed so as to facilitate the pursuit of these goals. (iii) In the past, the pattern of goals, and the body changed very gradually, possibly in step with a changing environment, in such a way that each small change enhanced reproductive success.

Two interpretations of Darwin's theory – or of neo-Darwinianism – need to be distinguished. They may be called the *anti-purposive* and the *purposive* interpretations.

Anti-purposivism interprets neo-Darwinianism in such a way that the theory helps us to eliminate purposiveness from Nature, the aim being to explain and understand the biological world in non-purposive terms, in terms of molecular biology, and ultimately in terms of the purposeless laws of chemistry and physics. The aim is to explain and understand ostensible purposiveness in the world by explaining it away, ultimately everything being explicable solely in physicalistic terms.

Purposivism, in contrast, interprets neo-Darwinianism in such a way that the theory enables us to explain and understand how and why purposiveness has evolved in Nature, in a way that is in accordance with physicalism. The task of purposive neo-Darwinianism is to enable us to explain and understand how the diverse purposive patterns exhibited by, and embodied in, plant and animal life, have gradually come to be superimposed upon the fixed pattern of physicalistic law. Purposivism accepts physicalism but is anti-reductionist.

Insofar as we understand ourselves as purposive beings (at the very least) anti-purposivism creates an entirely artificial (and thoroughly non-Darwinian) hiatus between the purposeless biological world of Nature, and the purposeful human world of history and the present.[2] This hiatus is automatically avoided by purposivism, the interpretation adopted here.

The generalized Darwinian research programme accepts physicalism, and seeks to understand how and why all purposiveness has evolved in the universe – especially purposiveness associated with what we value most in human life, such as sentience, consciousness, person-to-person understanding, science, art, freedom, love. This programme of research brings together, into a coherent field of inquiry, aspects of such diverse fields of research as orthodox Darwinian theory (given its purposive interpretation), the study of animal behaviour, palaeontology, archaeology, history, anthropology, psycho-neurology, artificial intelligence, psychology, sociology, philosophy, linguistics, semantics, history and philosophy of science, and history and philosophy of inquiry more generally (the history and philosophy of ideas and culture). Person-to-person understanding of people in the past is embedded in a more general animal-to-animal understanding (so brilliantly displayed by Jane Goodall, for example, in her almost 'anthropological' studies of chimpanzees). Animal-to-animal understanding

---

[2] Monod and Dawkins both incline towards adoption of the anti-purposive interpretation of Darwinian theory: as a result, both hold that evolution of a new kind comes into existence with the cultural evolution of humanity. See Monod (1974, chs. 8 and 9); Dawkins (1978, ch. 11).

involves not only endeavouring imaginatively to enter into the lives and experiences of animals: it also involves interpreting ourselves as animals – as close cousins of chimpanzees, for example. Animal-to-animal understanding is in turn embedded in the more general purposive understanding, this in turn being embedded in physicalistic understanding (which, however, is itself an evolution of person-to-person understanding).

In line with physicalism, this programme of research presupposes that goal-pursuing entities do not come abruptly into existence from prior purposeless states of affairs. There is no sponataneous generation of life. Furthermore, there is no abrupt initiation of new goal-directed activity, radically different from antecedent goal-directed activity, to such an extent that the new goal-directed activity is as inexplicable as spontaneous generation of life. All new goal-directed activity (it is presumed) can be explained and understood as arising as a slight, intelligible modification of prior goal-directed activity. Where radically new goal-directed activity does genuinely arise, this is due to an already existing capacity for innovation, creativity, originality or learning, gradually and intelligibly developed in the past and suddenly given the opportunity to flourish in a new way by a small, intelligible change of circumstances. Of course, there are a multitude of goal-directed activities going on in the world, associated especially with human life, that seem radically different from previous activities. It is these innovative activities that pose the problems that the generalized Darwinian programme seeks to solve.

One important general problem confronting this programme is the problem of how purposive beings create new purposive beings. Four possibilities are (a) exclusively genetic or biological reproduction; (b) genetic plus educational or cultural reproduction; (c) manufacture; (d) manufacture plus education ((c) and (d) arising in connection with robots).

A momentous development in evolution is the transition from (a) to (b). It is this which makes cultural evolution possible – the evolution of new ways of life even in the absence of genetic evolution. New kinds of actions, initiated by individual animals, are learned by offspring, culturally inherited as it were, and progressively developed during the course of a number of generations. Social and cultural changes that have taken place throughout human history, and more recent scientific, technological and associated social and cultural changes – unprecedented in their radical character and ever accelerating rapidity of occurrence when put into the context of biological evolution as a whole – both exemplify, and depend upon the prior existence of, cultural evolution. Much that is essential to our humanity, to our identity as the individual persons we now are, such as language, personal relationships, customs, institutions, values, exist and persist because of a long prior process of cultural evolution.

How does cultural reproduction and evolution itself gradually evolve from almost exclusively genetic reproduction and evolution? In order for it to be possible for animals to reproduce and evolve culturally it is essential for animals to possess two capacities: (i) the capacity to learn individually, and (ii) the capacity to imitate (itself, perhaps, a special kind of learning). It seems likely that the development of cultural reproduction is, in addition, associated with the development of parental care. For it is primarily when offspring are cared for by parents for some time that learning through imitating others is likely to have survival value. We may postulate, then, the gradual development of (i) the capacity to learn, (ii) the capacity to care for young, and (iii) the capacity to imitate, by means of almost exclusively genetic evolution. Parental care, for example, begins with care being taken to place eggs advantageously: this leads to guarding eggs; to moving and guarding newly-hatched offspring (performed by crocodiles), to feeding offspring (birds). When to what crocodiles and birds do there is added training in how to find food, hunt, or escape from predators – performed by many mammals – the conditions for cultural reproduction to occur are satisfied. In such conditions, mutations promoting the capacity to imitate and to learn from parental actions in youth will have survival value. Such mutations make cultural reproduction and evolution possible.

Whether a way of life is reproduced in an exclusively genetic way, or in a way that is in part genetic, in part cultural, is something that can in principle be determined empirically. Spiders spin webs and execute other aspects of a spider way of life entirely successfully even if reared in isolation from other spiders: here the way of life is passed on from body to body in an exclusively genetic way (in the given environment). In the case of many mammals, however, and especially the primates, this is not the case at all. Even if given the opportunity to survive and to learn how to survive in an isolated but otherwise carefully controlled environment, many mammals will, in these circumstances fail to develop the capacity to survive and reproduce if returned to their natural habitat. Young chimpanzees die simply from being deprived of the presence of their mothers.

The development of (b) genetic-plus-cultural-reproduction

(from prior (a) exclusively-genetic-reproduction) changes profoundly the character of evolution. In particular, it makes it possible for non-genetic, exclusively cultural changes in an animal's way of life to be an essential part of the cause of subsequent morphological changes of descendants, changes that are genetically reproduced, as Hardy (1965) especially has emphasized.

The great advantage of the generalized Darwinian research programme, just outlined, is that it provides a framework for understanding the deeds, achievements and experiences of people in a way that is compatible with the kind of knowledge and understanding achieved in the physical sciences, without being reducible to such knowledge and understanding. It promises to enable us to understand ourselves as a part of the biological domain without our humanity, our distinctive human value, being in any way denied: persons are not reduced to animals, and nor are animals misconceived to be persons. It holds out the hope that we can come to understand the human world as an integral part of the natural world without the meaning and value of the human world being thereby conceptually annihilated. The programme specifies in general terms what we must seek to do in order to develop a coherent understanding of nature and of ourselves which does justice to the character of both.

From the standpoint of the philosophy of wisdom, of course, this programme of research provides no more than a background to the central task of rational inquiry: to help us develop our overall goal of seeking reproductive success, inherited from our evolutionary past, so that it becomes the goal of living life lovingly, cooperatively helping to develop a less suffering, more loving human world.

My claim is that the above discussion shows how physicalism and experiential realism can be reconciled, in an intellectually fruitful way. But how, it may be asked, can *free will* be reconciled with physicalism?

A major part of the problem here is to arrive at an acceptable definition of 'free will'. I suggest that an acceptable definition must be such that it is clear that 'free will', in the defined sense, is something that is of great value to possess, the more valuable the better the definition.

An important part of what we ought to mean by free will, or freedom, can, I suggest, be put like this: to be free is to have the capacity (and the opportunity) to realize what is of value in life. We are free to the extent that we do, or do not, possess this capacity (and opportunity). Clearly, it is of great value to have 'free will' in this sense.

Granted that this conception of free will is accepted, then the above discussion, in showing how it is possible for there to be purposive human life of value immersed in a physicalistic universe, also shows how it is possible for there to be some degree of freedom associated with human life even though physicalism is true.

It deserves to be noted that freedom, in this sense, satisfies the Darwinian requirement of being something that can be understood to have developed *gradually*, in small steps, during the course of evolution. It develops gradually with the gradual development of the capacity to learn, to imitate, to dream and to imagine, to be sentient and conscious, and to be able to communicate (all of which exists in chimpanzees, for example).

How free are we? From the standpoint of this book, our freedom is to be judged in terms of our capacity and opportunity to avoid suffering and death and live life lovingly. Clearly, when judged from this perspective, human freedom is severely restricted.

In order to increase our freedom, in this sense, we need, quite generally, to improve our aims and methods as we live in such a way that we realize what is of value to us. Rational inquiry, pursued in accordance with the philosophy of wisdom, has as its basic task, to increase freedom!

An argument in support of the contention that mankind does indeed have the capacity to be free emerges from just that which seems to threaten the possibility of freedom – namely the success of theoretical physics (or natural philosophy). The argument can be put like this. Suppose physicalism is true. Suppose, that is, that the universe really is comprehensible in the kind of way modern physics holds it to be. In this case one cooperative human endeavour of great value has been extraordinarily successful, namely the endeavour of improving our knowledge and understanding of the universe. Here then is a practical demonstration of human freedom (as defined above). If physicalism is true, in short, mankind definitely does have the capacity to be free. *The truth of physicalism, far from threatening, actually serves to establish, the reality of human freedom.*

The argument of this chapter might be summarized as follows.

Two important lessons are to be learned from the success of physical science: a view of the world, and a methodology. The view of the world is *physicalism*, qualified by *experiential realism*. The methodology is *aim-oriented empiricism* generalized to become *aim-oriented rationalism*. The *generalized Darwinian research programme* holds out the hope of enabling us to improve our understanding of how cooperative aim-oriented rationalistic life of value might come to be in the physical universe. At the same time it emphasizes the fundamental importance of endeavouring to put cooperative aim-oriented rationalism into practice in our lives, within the framework indicated, so that we may come to develop a less cruel, more loving world.

CHAPTER ELEVEN

The Revolution is Under Way

At present standard empiricism and the philosophy of knowledge predominate in science, and in universities, in a very obvious way, as chapter 6 makes clear. From this it might be concluded that there are few signs of change in academic inquiry, from knowledge to wisdom, and little hope that such a change will come to be in the foreseeable future, however urgent the need may be, and however decisive the reasons may be, for making such a change.

This gloomy conclusion is, I believe, a mistake. There is a growing groundswell of opinion and effort already devoted, in various ways, to bringing about changes in science, technology, scholarship, education, medicine, welfare, aid, politics, the media and elsewhere that can be interpreted as pioneer attempts to implement aspects of what has here been called 'the philosophy of wisdom'. These diverse efforts are, however, scattered and isolated. Individuals find themselves battling alone against general incomprehension and misrepresentation. Those concerned to develop academic psychology along rather more philosophy-of-wisdom lines are perhaps unaware of similar efforts being made in sociology, economics, philosophy, or education. There is a general failure to appreciate the need for a coordinated and comprehensive change in intellectual aims and methods throughout all of academic inquiry and education. Above all, the current prevalence of standard empiricist and philosophy-of-knowledge intellectual standards ensures that these efforts do not receive the attention, discussion, and publicity that they deserve. Intellectual standards inevitably, and quite properly, function as a form of censorship. Standard empiricism and the philosophy of knowledge are no exception. At present potentially excellent contributions to inquiry from the standpoint of the philosophy of wisdom do not get published – or when published do not get noticed – just because of a failure to conform to the edicts of standard empiricism and the philosophy of knowledge. Academics are discouraged from giving intellectual priority in their work to the tasks of articulating problems of living, proposing and criticizing possible solutions, as they know full well that such work, however urgently needed and intellectually excellent, will not be accepted for publication in that it will not amount to potential 'contributions to knowledge'. In these ways the institutional illusion is created that academics universally accept the philosophy of knowledge, even though in fact there are many who hold that the currently adopted intellectual system – its aims, problems, priorities and values – is profoundly and damagingly defective.

I am myself well aware of just how potently the philosophy of knowledge, as a result of being built into the institutional structure of academic inquiry, operates to censor out of existence work that fails to conform to its edicts. I have encountered this again and again in my own work – above all in my attempts, during the past ten to twenty years, to communicate and publish the proposals and arguments of this book!

The themes of this book have their origins, for me, in my childhood. For as long as I can remember I have had the passionate desire to get to the bottom of things, to understand. Probably for all too human reasons, I have wanted to discover 'the secret of the universe', the riddle of life. To begin with this took the form of a desire to understand the ultimate structure of the physical universe. As a twelve-year-old, I read with fascinated incomprehension accounts of nuclear physics to be found in *Penguin Science News*, and Eddington and Russell on relativity and quantum theory. It was above all the mystery, the incomprehensibility, of this strange world of physics that appealed to my imagination. Here was this extraordinary world – of time slowing down and space contracting, of curved space-time, of particles that are also waves, of almost infinitely vast galaxies and infinitely minute atoms – apparently so different from the familiar world of stones, trees and people: and yet it was this other mysterious world that was the real world, the common-sense world being largely an illusion. With the customary unselfconscious audacity of the young, I decided that I would discover the secret of all this mystery, and thus reveal to the world the true meaning of existence.

With the onset of adolescence, however, I discovered literature. I plunged into the vivid, dramatic, extraordinary worlds of Dostoevsky, Kafka, Virginia Woolf, Chekhov, Stendhal, Shelley,

Hazlett, Fielding, D.H. Lawrence, Rex Warner, Emily Brontë. Here, I began to feel was reality: the vivid, dramatic and extraordinary inner world of human life, the inner world of imaginative experience. I decided I would discover the innermost secret of this mysterious and passionate world of human experience by writing novels. I would create a living and breathing universe, so real in its dramatic intensity that it would all but engulf the real world.

My attempts to do this failed miserably. A failed theoretical physicist, mathematician and novelist, I decided, after a spell of national service, to return to university to do philosophy. My efforts 'at being a genius' (as I thought of it then) had failed, and were obviously absurd in any case. Without any great expectations – in order merely to indulge an interest in trying to understand how things fit together – I became an undergraduate at Manchester University.

But after a year I made what seemed to me then a great discovery. As I put it then: 'the riddle of the universe is the riddle of our desires'. Philosophy devotes itself to the problems of knowledge, thus presuming, without question, that the basic aim of inquiry is to acquire knowledge. It is this presumption that is the mistake. The proper basic aim of philosophy is to help us resolve the riddle of our desires. It is not in the ultimate nature of the universe, nor in the ultimate nature of our inner life, that the answer to the riddle of life lies; it lies rather in what might be termed the region of overlap between the two – in the familiar miracle of this experienced world (its familiarity all too often, alas, dimming our perception of its miraculousness).

I had sought the answer to the riddle of life in the ultimate nature of the physical universe, and in the ultimate nature of our inner world. Actually the answer to the riddle of my life lay around me all the time, in the experience of living my life. I had striven to be a physicist and a novelist and had neglected to be what was for me the thing of most value, myself. The miracle upon miracle is this moment-by-moment experience of living, the outcome of the interaction between our unknown inner world and the unknown outer world. In discovering and participating in this experienced world of colour, sound, landscapes, people, beauty, we all exhibit passionate intellectual and imaginative resources that far outstrip what is revealed in the work of the greatest scientists and artists: but over-familiarity and misconceptions conspire to make us lose sight of the miraculous character of the worlds we inhabit. I wrote down these 'discoveries' of mine, submitted the manuscript for publication, and looked forward to telling the philosophers of Manchester University about the new territories for philosophy that I had stumbled across.

The manuscript was rejected; and I found that in seminars and lectures I could scarcely open my mouth. I became increasingly horrified by academic philosophy; to me, it seemed to be either totalitarian in motivation, or utterly trivial. Great philosophers did not offer their intellectual visions of reality in an intellectually honourable way as *possibilities* – more explicitly articulated and scrutinized versions of the philosophies of reality we all create and discard casually, as we live. On the contrary, from Plato to Wittgenstein, they sought to *prove* the final and complete truth of their personal vision, thus in effect, as far as I was concerned, endeavouring to set up a sort of intellectual dictatorship, all other minds and lives to be faithful copies of their own, programmed by indoctrination masquerading as education. Apart from this, there was the triviality of ordinary language philosophy and conceptual analysis. No one seemed to be interested in the obvious and important endeavour of imaginatively articulating and scrutinizing the basic, problematic aims of life (presumably because academic philosophers never questioned their own aims). I began to suspect I was living in a new dark age.

Then I discovered Karl Popper, and especially *The Open Society and its Enemies*, and I heaved a sigh of relief. Here was a work passionately concerned with a profoundly important problem: how are we to achieve civilization? What are the basic *problems*, and what role does reason play in helping us to solve these problems? Popper had tackled these problems in a wholly responsible way intellectually and morally, and with a wealth of detailed scholarship. The covertly totalitarian character of philosophy was, with the discussion of Plato, brilliantly unmasked. With Popper's work, I concluded, the basic problems of epistemology, methodology, political and social philosophy had received their definitive solution.

Subsequently, however, as a result of pondering difficulties associated with Popper's claim to have solved the problem of induction (discussed in chapter 9), I came to the conclusion that the source of the trouble was that Popper, along with almost all scientists, seriously misrepresented the basic intellectual aim of science. The argument of chapter 5 unfolded itself before my eyes. My discovery of ten years earlier had re-emerged as the idea that all rational inquiry – and not just philosophy – should have as its basic task to help us to improve our aims and methods as we live so that we may realize what is of most value to us in life. I put pen to paper, and before long discovered that now that I had something important to communicate publication seemed to be all but impossible.

I am quite sure that many others have followed lines of argument not so dissimilar from those spelled out in this book. Doubtless they too have encountered what I have encountered: that the philosophy of knowledge, as a result of being already institutionalized, tends to block both criticism of itself and attempts to pursue inquiry along rather more philosophy-of-wisdom lines. There is a further point. It is at present especially difficult for people without academic qualifications to speak up and be heard within academic contexts. During ten years of advocating the philosophy of wisdom to all and sundry, I have found that most non-academic women know what I am talking about straightaway, most academic men do not, with non-academic men and academic women falling somewhere in between.

There can be no doubt that during the last ten to fifteen years a multitude of developments have taken place within and without the academic world that can be interpreted as disparate attempts to put into practice inquiry corresponding to what has here been called 'the philosophy of wisdom'. Here is an impressionistic indication of some of these developments – with the emphasis on those that have occurred in Britain.

During the period in question, there has been an enormous increase in concern about environmental and ecological problems: problems of pollution, depletion of finite natural resources, destruction of plant and animal life, growth of the world population. A number of books come immediately to mind: Rachel Carson's *Silent Spring* (1962); Barry Commoner's *Science and Survival* (1966); The Club of Rome's report; Meadows *et al.*, *The Limits to Growth* (1972); Dubos and Ward, *Only One Earth* (1972); Maddox, *The Doomsday Syndrome* (1972); Goldsmith *et al.*, 'A blueprint for survival' (1972); Foley, *The Energy Question* (1976); Allaby, *Inventing Tomorrow* (1976); Ward, *Progress for a Small Planet* (1979); Allen, *How to Save the World* (1980); Eckholm, *Down to Earth* (1982). Groups such as the Friends of the Earth have had an effect in increasing public awareness of environmental issues. The problems have been increasingly discussed in the media. The *Ecologist* has published articles on themes closely related to that of this book (see, for example, Skolinowski, 1975, and the issue devoted to the 'Scientific straightjacket', 11, (1), 1981). Political parties have even been formed around ecological issues, with West Germany's Green Party winning seats in the Bundestag as I write.

There has been an increasing awareness of the plight of people living in the third world, as evinced by such things as Victor Zorza's 'Village Voice' column and the Third World Review, both in the *Guardian*, the rise of journals like *New Internationalist* and the publication of such books as George, *How the Other Half Dies* (1976); P. Harrison, *Inside the Third World* (1979); Brandt *et al.*, *North-South: A Programme for Survival* (1980). Some authors have attempted to give a comprehensive survey of the most important global problems that confront mankind in the decades to come; for example Heilbroner, *An Inquiry into the Human Prospect* (1975); and most notably Higgins, *The Seventh Enemy* (1978).

Closely associated with these concerns, movements have arisen seeking to develop and promote alternative, intermediate and appropriate technology; see, for example Schumacher, *Small is Beautiful* (1973); Dickson, *Alternative Technology* (1974); Cooley, *Architect or Bee?* (1980). I have mentioned already renewed interest in Britain in cooperatives; in addition, see *In the Making: A Directory of Radical Cooperation* (1981).

There is also the movement for social responsibility of science, promoted for example by the British Society for Social Responsibility in Science, and by the society's journal *Science for the People*. Initially this movement began outside, or on the fringes of, universities and colleges of technology. Subsequently, it has had a considerable impact on courses and departments in universities and colleges of technology. Science in a Social Context (SISCON), guided by the wise stewardship of Dr Bill Williams at Leeds University, produced during the seventies over forty booklets designed to provide background material for courses in higher education on issues having to do with science, technology and society. Over twelve universities and colleges of technology in Britain now have departments or give courses devoted to such issues (although in 1982 some began to face severe difficulties due to cut-backs in expenditure on higher education). Analogous but much more wide-ranging and radical developments have taken place in the USA, as Heitowit (1977) shows. The UNESCO publication *World Directory of Research Projects, Studies and Courses in Science and Technology Policy* (1981a) list over 1,000 departments, institutes or units devoted to such issues. (See also UNESCO, 1981b).

Criticisms of modern science, technology and industrial society, conducted initially primarily from outside the scientific world by such writers as Jacques Barzun, Jacques Ellul, Theodore Roszak, and Ivan Illich, have gradually had a certain impact. Nowadays there are many distinguished scientists, Nobel prizewinners, Fellows of the Royal Society, pillars of the establishment, who are profoundly disturbed and concerned by the priorities of current scientific and technological research and by the use to which such research is put. Something of this can for example be detected in Dyson's recent scientific autobiography *Disturbing the Universe* (1981). It is to be found in the *Bulletin of the Atomic Scientists*, and in recent literature on the threat of the bomb, such as Ryle *Towards the Holocaust* (1981) and Rotblat (ed.) *Scientists, the Arms Race and Disarmament* (1983). And it is to be found in Jerry Ravetz's call for a more humanitarian and critical science, see his *Scientific Knowledge and its Social Problems* (1971).

Psychiatrists, emerging from a post-Freudian background, have kept alive a tradition of giving intellectual priority to problems of living we encounter in seeking that which is of value to us. In this context, most notable are the numerous publications of Erich Fromm, such as *The Fear of Freedom* (1942), *The Sane Society* (1963), and *To Have or to Be?* (1979). There are also such works as Rollo May, *Love and Will* (1972); Axline, *Dibs: In Search of Self* (1971); Szasz, *The Myth of Mental Illness* (1961) and Laing, *The Divided Self* (1965). There is also Sacks' remarkable book *Awakenings* (1976) about people suffering from severe forms of Parkinson's disease. The book does full justice to the primacy of the problems of living suffered by the people in question, without in any way denying or blurring the physiological aspects of these problems.

Finally, there are eight books I wish to refer to which, in very different ways, pursue themes related to those of the present book. First Peter Gay's *The Enlightenment: An Interpretation* (1973), a magnificent evocation of the work and thought of the 'philosophes' of the eighteenth century, passionately devoted as they were to the progressive achievement of enlightenment through critical reason. Academics today might well regard the life-work of 'philosophes' like Voltaire and Diderot as paradigmatic of what academic work ought to be. Second, there is Brian Easlea's *Liberation and the Aims of Science* (1973), a serious and heartfelt exploration of problems concerning the aims of science, and how these might be transformed so that science offers more help with the task of building a less suffering, more loving world.

Third, there is Robert Pirsig's *Zen and the Art of Motorcycle Maintenance* (1974), brilliantly exploring, partly in fictional or autobiographical form, themes closely related to those of this book. Fourth, there is Iris Murdoch's *The Sovereignty of Good* (1970), in which I seemed to find depicted something close to aim-oriented rationalism. Fifth, there is Mary Midgley's *Beast and Man* (1978), a book which has many interesting things to say about the problems of how our humanity has arisen from, and is related to, our animal past. Sixth, there is John Kekes' *The Nature of Philosophy* (1980), in which it is argued that 'it is the task of philosophy to show how to live well by the construction and rational justification of worldviews' (p. xii). Seventh, there is David Collingridge's *The Social Control of Technology* (1981), a book which has detailed, incisive and important things to say about 'one of the most pressing problems of our time – "can we control our technology – can we get it to do what we want and can we avoid its unwelcome consequences?" ' (p. 11). Finally, there is Colin Norman's *The God that Limps* (1981), already referred to, which considers the extent to which priorities of scientific and technological research succeed and fail to correspond to human need, in a global context.

The intellectual revolution, from knowledge to wisdom, is already under way. It will need, however, much wider cooperative support – from scientists, scholars, students, research councils, university administrators, vice chancellors, teachers, the media and the general public – if it is to become anything more than what it is at present, a fragmentary and often impotent movement of protest and opposition, often at odds with itself, exercising little influence on the main body of academic work.

Bibliography

Ackermann, R. (1961) 'Inductive simplicity', *Philosophy of Science*, 28, pp. 152–61.

Allaby, M. (1977) *Inventing Tomorrow*, Abacus, London.

Allen, R. (1980) *How to Save the World: Strategy for World Conservation*, Kogan Page, London.

Attenborough, D. (1981) *Life on Earth*, Fontana, London.

Axline, V. (1971) *Dibs: In Search of Self*, Penguin, Harmondsworth.

Bacon, F. (1975) *The New Organon*, ed. F. Anderson, Bobbs-Merrill, Indianapolis (first published 1620).

Bacon, F. (1980) *The Advancement of Learning and New Atlantis*, ed. A. Johnston, Clarendon Press, Oxford (first published 1605).

Bannister, D. and Fransella, F. (1971) *Inquiring Man*, Penguin, Harmondsworth.

Barker, S.F. (1961) 'On simplicity in empirical hypotheses', *Philosophy of Science*, 28, pp. 162–71.

Barnes, B. (1974) *Scientific Knowledge and Sociological Theory*, Routledge & Kegan Paul, London.

Barnet, R. (1972) *Intervention and Revolution*, Paladin, London.

Barzun, J. (1964) *Science: The Glorious Entertainment*, Secker and Warburg, London.

Bauman, Z. (1978) *Hermeneutics and Social Science*, Hutchinson, London.

Bell, D. and Kristal, I. (eds) (1981) *The Crisis in Economic Theory*, Basic Books, New York.

Benacerraf, P. and Putnam, H. (eds) (1964) *Philosophy of Mathematics: Selected Readings*, Prentice-Hall, Englewood Cliffs.

Ben-David, J. (1971) *The Scientist's Role in Society*, Prentice-Hall, Englewood Cliffs.

Berlin, I. (1976) *Vico and Herder*, Hogarth Press, London.

Berlin, I. (1979) *Against the Current*, Hogarth Press, London.

Berlin, I. (1980) *Concepts and Categories, Oxford University Press*.

Bernal, J.D. (1967) *The Social Function of Science*, MIT Press (first published 1939).

Blaug, M. (1968) *Economic Theory in Retrospect*, Irwin, Illinois.

Blaug, M. (1980) *The Methodology of Economics*, Cambridge University Press.

Bleicher, J. (1980) *Contemporary Hermeneutics*, Routledge & Kegan Paul, London.

Bloor, D. (1976) *Knowledge and Social Imagery*, Routledge & Kegan Paul, London.

Bolzano, E. (1972) *Theory of Science*, Blackwell, Oxford (first published 1837).

Borger, R. and Cioffi, F. (eds) (1970) *Explanation in the Behavioural Sciences*, Cambridge University Press.

Brandt, W. *et al.* (1980) *North-South: A Programme for Survival*, Pan Books, London.

Broadbent, D.E. (1973) *In Defence of Empirical Psychology*, Methuen, London.

Brodbeck, M. (ed.) (1968) *Readings in the Philosophy of the Social Sciences*, Macmillan, New York.

Bronowski, J. (1956) *Science and Human Values*, Meisner, New York.

Broom, L. *et al.* (1981) *Sociology*, Harper and Row, New York.

Brown, P. (ed.) (1973) *Radical Psychology*, Tavistock, London.

Brown, S. *et al.* (1981) *Conceptions of Inquiry*, Meuthen, London.

Bunge, M. (1961) 'The weight of simplicity in the construction and assaying of scientific theories', *Philosophy of Science*, 28, pp. 120–49.

Bunge, M. (ed.) (1967) *Quantum Theory and Reality*, Springer, Berlin.

Burks, A.W. (1977) *Chance, Cause and Reason*, Chicago University Press.

Burtt, E.A. (1965) *In Search of Philosophical Understanding*, New American Library, New York.

Cairns, D. and Dressler, D. (1973) *Sociology*, Knopf, New York.

Calder, N. (1981) *Nuclear Nightmares*, Penguin, Harmondsworth.

Carson, R. (1972) *Silent Spring*, Penguin, Harmondsworth (first published 1962).

Chain, E. (1970) 'Social responsibility and the scientist', *New Scientist*, 22 October, p. 166.

Chalmers, A.F. (1976) *What is this Thing Called Science?*, Open University Press, Milton Keynes.

Cioffi, F. (1970) 'Freud and the idea of a pseudo science', in Borger and Coiffi (eds) (1970) pp. 471–99.

Collingridge, D. (1981) *The Social Control of Technology*, Open University Press, Milton Keynes.

Colodny, R. (ed.) (1965) *Beyond the Edge of Certainty*, Prentice-Hall, New York.

The Committee . . . (1981) *Hiroshima and Nagasaki: the Physical, Medical, and Social Effects of the Atomic Bombings*, Hutchinson, London.

Commoner, B. (1966) *Science and Survival*, Gollancz, London.

Cooley, M. (1980) *Architect or Bee?*, Langley Technical Services, Slough.

Corwin, R. and Nagi, S. (1972) *The Social Contexts of Research*, Wiley-Interscience, London.

Davies, J.T. (1973) *The Scientific Approach*, Academic Press, London.

Davies, P. (1979) *The Forces of Nature*, Cambridge University Press.

Dawkins, R. (1978) *The Selfish Gene*, Paladin, London.

de Bono, E. (1972) *Children Solve Problems*, Penguin, Harmondsworth.

de Bono, E. (1974) *The Use of Lateral Thinking*, Penguin, Harmondsworth.

Dickson, D. (1974) *Alternative Technology*, Fontana, London.

Dubos, R. and Ward, B. (1972) *Only One Earth*, Penguin, Harmondsworth.

Duhem, P.(1962) *The Aim and Structure of Physical Theory*, Atheneum, New York (first published 1905).

Dukas, H. and Hoffmann, B. (1979) *Albert Einstein: the Human Side*, Princeton University Press.

Dyal, J.A. *et al.* (1975) *Readings in Psychology: the Search for Alternatives*, McGraw-Hill, New York.

Dyson, F. (1981) *Disturbing the Universe*, Pan, London.

Easlea, B. (1973) *Liberation and the Aims of Science*, Chatto and Windus, London.

Eckholm, E. (1982) *Down to Earth: Environment and Human Needs*, Pluto Press, London.

Einstein, A. (1953) 'The laws of science and the laws of ethics' in H. Feigl and M. Brodbeck (eds) *Readings in the Philosophy of Science*, Appleton Century Crofts, New York, p. 779.

Einstein, A. (1973) *Ideas and Opinions*, Souvenir Press, London (first published 1954).

Elliot, J. (1978) *Conflict or Cooperation? The Growth of Industrial Democracy*, Kogan Page, London.

Ellul, J. (1964) *The Technological Society*, Vintage Books, New York.

Encounter (1980) 'What's wrong with the Brandt Report?' 55 (6), pp. 12–30.

Eysenck, H.J. (1965) *Fact and Fiction in Psychology*, Penguin, Harmondsworth.

Feyerabènd, P. (1965) 'Problems of empiricism' in R. Colondy (ed.) (1965), pp. 145–260.

Feyerabend, P. (1975) *Against Method: Outline of an Anarchistic Theory of Knowledge*, New Left Books, London.

Feyerabend, P. (1978) *Science in a Free Society*, New Left Books, London.

Foley, G. (1981) *The Energy Question*, Penguin, Harmondsworth.

Fonda, L. *et al.* (1978) 'Decay theory of unstable quantum systems', *Reports on Progress in Physics*, 41, pp. 587–631.

Friedman, M. (1968) 'The methodology of positive economics', in Brodbeck (1968), pp. 508–28.

Frisch, M. (1974) *Homo Faber*, Penguin, Harmondsworth.

Fromm, E. (1942) *The Fear of Freedom*, Kegan Paul, London.

Fromm, E. (1963) *The Sane Society*, Routledge & Kegan Paul, London.

Fromm, E. (1979) *To Have or to Be?* , Abacus, London.

Gaa, J. *et al.* (1977) 'Value issues in science, technology and medicine', *Philosophy of Science*, 44, pp. 511–618.

Gay, P. (1973) *The Enlightenment: an Interpretation*, Wildwood House, London.

Gellner, E. (1959) *Words and Things*, Gollancz, London.

George, S. (1976) *How the Other Half Dies*, Penguin, Harmondsworth.

Giddens, A. (1976) *New Rules of Sociological Method*, Hutchinson, London.

Giner, S. (1972) *Sociology*, Martin Robertson, London.

Goffman, E. (1971) *The Presentation of Self in Everyday Life*, Penguin, Harmondsworth.

Goldsmith, E. *et al.* (1972) 'A blueprint for survival', *The Ecologist*, 2 (1), pp. 1–43.

Goodall, J. (1971) *In the Shadow of Man*, Collins, London.

Goodman, N. (1961) 'Safety, strength, simplicity', *Philosophy of Science*, 28, pp. 150–1.

Goodman, N. (1972) *Problems and Projects*, Bobbs-Merrill, New York.

Goodwin, P. (1982) *Nuclear War: the Facts*, Macmillan, London.

Gouldner, A. (1970) *The Coming Crisis in Western Sociology*, Heinemann, London.

Graham, L. (1981) *Between Science and Values*, Columbia University Press, New York.

Greenberg, D.S. (1971) *The Politics of Pure Science*, New American Library (first published 1967).

Habermas, J. (1972) *Knowledge and Human Interests*, Heinemann, London.

Hadamard, J. (1954) *The Psychology of Invention in the Mathematical Field*, Dover, New York.

Hagstrom, W.O. (1965) *The Scientific Community*, Basic Books, New York.

Hardy, A. (1965) *The Living Stream*, Collins, London.

Harris, J. (1974) 'Popper's definitions of "verisimilitude" ', *British Journal for the Philosophy of Science*, 25, pp. 160–6.

Harris, M. (1979) *Cultural Materialism*, Random House, New York.

Harris, N. (1968) *Beliefs in Society*, Penguin, Harmondsworth.

Harrison, P. (1979) *Inside the Third World*, Penguin, Harmondsworth.

Harrison, R. (1979) *Rational Action*, Cambridge University Press.

Hayek, F. (1967) *Studies in Philosophy, Politics and Economics*, Chicago University Press.

Heather, N. (1976) *Radical Perspectives in Psychology*, Methuen, London.

Heilbroner, R. (1975) *An Inquiry into the Human Prospect*, Calder and Boyars, London.

Heitowit, E. (1977) *Science, Technology and Society*, Cornell University, Ithaca.

Hempel, C.G. (1965) *Aspects of Scientific Explanation*, Collier-Macmillan, London.

Herschel, J.F.W. (1831) *Preliminary Discourse on the Study of Natural Philosophy*, Longman, London.

Hesse, M. (1974) *The Structure of Scientific Inference*, Macmillan, London.

Higgins, R. (1978) *The Seventh Enemy: the Human Factor in the Global Crisis*, Hodder and Stoughton, London.

Hoffmann, B. (1972) *Albert Einstein: Creator and Rebel*, Hart-Davis, London.

Hollis, M. and Nell, E. (1975) *Rational Economic Man*, Cambridge University Press.

Holton, G. (1973) *Thematic Origins of Scientific Thought: Kepler to Einstein*, Harvard University Press.

Hume, D. (1959) *A Treatise of Human Nature*, Everyman, London (first published 1738).

Hutchison, T.W. (1977) *Knowledge and Ignorance in Economics*, Blackwell, Oxford.

Hutchison, T.W. (1978) *Revolutions and Progress in Economic Knowledge*, Cambridge University Press.

Huxley, A. (1977) 'Evidence, clues and motives in science', *Times Higher Educational Supplement*, 2 September.

In the Making: A Directory of Radical Co-operation (1981), ITM, Sutton.

Jammer, M. (1966) *The Conceptual Development of Quantum Mechanics*, McGraw-Hill, New York.

Jaspers, K. (1960) *The Idea of the University*, Peter Owen, London.

Jeffrey, R.C. (1965) *The Logic of Decision*, McGraw-Hill, New York.

Jevons, W.S. (1924) *The Principles of Science*, Macmillan, London.

Joynson, R.B. (1970) 'The breakdown of modern psychology', *Bulletin of the British Psychological Society*, 23, pp. 261–9.

Jungk, R. (1960) *Brighter than a Thousand Suns*, Penguin, Harmondsworth.

Kant, I. (1961) *Critique of Pure Reason*, Macmillan, London (first published 1781).

Keat, R. and Urry, J. (1975) *Social Theory as Science*, Routledge & Kegan Paul, London.

Kekes, J. (1980) *The Nature of Philosophy*, Blackwell, Oxford.

Kelly, G. (1955) *The Psychology of Personal Constructs*, Norton, New York.

Keynes, J.N. (1890) *The Scope and Method of Political Economy*, Macmillan, London.

Köhler, W. (1927) *The Mentality of Apes*, Kegan Paul, London.

Koestler, A. (1964) *The Sleepwalkers*, Penguin, Harmondsworth.

Koestler, A. and Smithies, J.R. (eds) (1969) *Beyond Reductionism – the Alpbach Symposium*, Hutchinson, London.

Kuhn, T.S. (1962) *The Structure of Scientific Revolutions*, Chicago University Press.

Kuhn, T.S. (1970) 'Reflections on my critics', in Lakatos and Musgrave (1970), pp. 231–78.

Kuhn, T.S. (1977) *The Essential Tension*, Chicago University Press.

Kyburg, H. (1970) *Probability and Inductive Logic*, Collier-Macmillan, London.

Laing, R.D. (1965) *The Divided Self*, Penguin, Harmondsworth.

Lakatos, I. (1970) 'Falsification and the methodology of scientific research programmes', in Lakatos and Musgrave (1970), pp. 91–195.

Lakatos, I. (1976) *Proofs and Refutations*, Cambridge University Press.

Lakatos, I. and Musgrave, A. (eds) (1970) *Criticism and the Growth of Knowledge*, Cambridge University Press.

Laudan, L. (1977) *Progress and its Problems: Towards a Theory of Scientific Growth*, Routledge & Kegan Paul, London.

Locke, J. (1961) *An Essay Concerning Human Understanding*, Everyman, London (first published 1690).

Maddox, J. (1972) *The Doomsday Syndrome*, Macmillan, London.

Mandrou, R. (1978) *From Humanism to Science 1400–1700*, Penguin, Harmondsworth.

Manuel, F.E. (1968) *A Portrait of Newton*, Harvard University Press, Cambridge, Mass.

Marx, K. (1921) *Capital*, Kerr, Chicago.

Maxwell, N. (1966) 'Physics and common sense', *The British Journal for the Philosophy of Science*, 16, pp. 295–311.

Maxwell, N. (1968a) 'Can there be necessary connections between successive events?' *The British Journal for the Philosophy of Science*, 19, pp. 1–25.

Maxwell, N. (1968b) 'Understanding sensations', *Australasian Journal of Philosophy*, 46, pp. 127–46.

Maxwell, N. (1972a) 'A critique of Popper's views on scientific method', *Philosophy of Science*, 39, pp. 131–52.

Maxwell, N. (1972b) 'A new look at the quantum mechanical problem of measurement', *American Journal of Physics*, 40, pp. 1431–5.

Maxwell, N. (1974) 'The rationality of scientific discovery', *Philosophy of Science*, 41, pp. 123–53 and 247–95.

Maxwell, N. (1976a) 'Towards a micro-realistic version of quantum mechanics', *Foundations of Physics*, 6, pp. 275–92 and 661–76.

Maxwell, N. (1976b) *What's Wrong With Science? Towards a People's Rational Science of Delight and Compassion*, Bran's Head Books, Hayes, Middlesex.

Maxwell, N. (1979) 'Induction, simplicity and scientific progress', *Scientia*, 114, pp. 629–53.

Maxwell, N. (1980) 'Science, reason, knowledge and wisdom: a critique of specialism', *Inquiry*, 23, pp. 19–81.

Maxwell, N. (1982) 'Instead of particles and fields: a micro-realistic quantum "smearon" theory', *Foundations of Physics*, 12, pp. 607–31.

May, R. (1972) *Love and Will*, Fontana, London.

Meadows, D.H. *et al.* (1974) *The Limits to Growth*, Pan Books, London.

Medvedev, Z.A. (1969) *The Rise and Fall of T.P. Lysenko*, Columbia University Press, New York.

Medvedev, R. and Medvedev, Z.A. (1971) *A Question of Madness*, Macmillan, London.

Melden, A. (1960) *Ethical Theories*, Prentice-Hall, Englewood Cliffs.

Merton, R. (1970) *Science, Technology and Society in Seventeenth-Century England*, Harper and Row, New York (first published 1939).

Midgley, M. (1978) *Beast and Man*, Harvester Press, Sussex.

Mill, J.S. (1843) *System of Logic*, Parker, London.

Miller, D. (1974) 'Popper's qualitative theory of verisimilitude', *The British Journal for the Philosophy of Science*, 25, pp. 166–77.

Mises, L. von (1960) *Epistemological Problems of Economics*, Nostrand, Princeton.

Monod, J. (1974) *Chance and Necessity*, Fontana, London.

Morgenstern, O. and von Neumann, J. (1944) *Theory of Games and Economic Behavior*, Princeton University Press.

Mulkay, M. (1979) *Science and the Sociology of Knowledge*, George Allen and Unwin, London.

Mulrey, J. (ed.) (1981) *The Nature of Matter*, Clarendon Press, Oxford.

Murdoch, I. (1970) *The Sovereignty of Good*, Routledge & Kegan Paul, London.

Nagel, E. (1961) *The Structure of Science*, Routledge & Kegan Paul, London.

Newton, I. (1962) *Principia*, University of California Press, Berkeley (first published 1687).

Newton-Smith, W.H. (1982) *The Rationality of Science*, Routledge & Kegan Paul, London.

Nickles, T. (ed.) (1980) *Scientific Discovery, Logic and Rationality*, and *Scientific Discovery: Case Studies*, Reidel, Dordrecht.

Nisbet, R. (1971) *The Degradation of the Academic Dogma: the University in America 1945–1970*, Heinemann, London.

Norman, C. (1981) *The God that Limps*, Norton, New York.

Oakeshott, R. (1978) *The Case for Workers' Co-ops*, Routledge & Kegan Paul, London.

O'Connor, D.J. (1973) *Free Will*, Macmillan, London.

Outhwaite, W. (1975) *Understanding Social Life*, Allen & Unwin, London.

Pais, A. (1980) 'Einstein on particles, fields, and the quantum theory', in H. Woolf (ed.), *Some Strangeness in the Proportion*, Addison-Wesley, Reading, Mass., pp. 197–251.

Pais, A. (1982) *Subtle is the Lord . . .*, Clarendon Press, Oxford.

Passmore, J. (1978) *Science and its Critics*, Duckworth, London.

Peirce, C.S. (1931–58) *Collected Papers*, Harvard University Press.

Pirsig, R. (1974) *Zen and the Art of Motorcycle Maintenance*, Bodley Head, London.

Plotkin, H.C. and Odling-Smee, F.J. (1981) 'A multiple-level model of evolution and its implication for sociobiology', *The Behavioural and Brain Sciences*, 4, pp. 225–68.

Poincaré, H. (1952) *Science and Hypothesis*, Dover, New York (first published in English in 1905).

Polanyi, M. (1958) *Personal Knowledge*, Routledge & Kegan Paul, London.

Polya, G. (1957) *How to Solve It*, Anchor Books, New York.

Popper, K.R. (1959) *The Logic of Scientific Discovery*, Hutchinson, London (first published 1934).

Popper, K.R. (1961) *The Poverty of Historicism*, Routledge & Kegan Paul, London.

Popper, K.R. (1963) *Conjectures and Refutations*, Routledge & Kegan Paul, London.

Popper, K.R. (1967) 'Quantum mechanics without the "observer" ', in Bunge (1967), pp. 7–44.

Popper, K.R. (1969) *The Open Society and its Enemies*, Routledge & Kegan Paul, London (first published 1945).

Popper, K.R. (1972) *Objective Knowledge*, Oxford University Press.

Popper, K.R. (1976) 'The logic of the social sciences', in *The Positivist Dispute in German Sociology*, Heinemann, London.

Popper, K.R. (1976) 'The logic of the social sciences', in *The Positivist Verlag, London.

Przibram, K. (ed.) (1967) *Letters on Wave Mechanics*, Vision Press, London.

Quinton, A. (ed.) (1968) *Political Philosophy: Oxford Readings in Philosophy*, Oxford University Press.

Ravetz, J.R. (1971) *Scientific Knowledge and Its Social Problems*, Clarendon Press, Oxford.

Raz, J. (1975) *Practical Reason and Norms*, Hutchinson, London.

Raz, J. (ed.) (1978) *Practical Reasoning*, Oxford University Press.

Reichenbach, H. (1961) *Experience and Prediction*, Chicago University Press (first published 1938).

Rescher, N. (1965) 'The ethical dimension of scientific research', in R. Colodny (1965), pp. 261–76.

Robbins, L. (1952) *An Essay on the Nature and Significance of Economic Science*, Macmillan, London.

Robbins, L. (1980) *Higher Education Revisited*, Macmillan, London.

Robinson, J. (1960) *Essays in Economic Analysis*, Macmillan, London.

Rose, H. and Rose, S. (eds) (1976) *The Radicalization of Science*, Macmillan, London.

Roszak, T. (ed.) (1969) *The Dissenting Academy*, Penguin, Harmondsworth.

Roszak, T. (1970) *The Making of a Counter Culture*, Faber, London.

Rotblat, J. (1983) *Scientists, the Arms Race and Disarmament*, Taylor and Francis, London.

Rudner, R. (1953) 'The scientist qua scientist makes value judgements', *Philosophy of Science*, 20, pp. 1–6.

Rudner, R. (1961) 'An introduction to simplicity', *Philosophy of Science*, 28, pp. 109–19.

Russell, B. (1948) *Human Knowledge: Its Scope and Limits*, Allen and Unwin, London.

Russell, B. (1956) *Portraits from Memory*, Allen and Unwin, London.

Ryle, M. (1981) *Towards the Holocaust*, Menard Press, London.

Sacks, W.O. (1976) *Awakenings*, Penguin, Harmondsworth.

Sagan, C. (1978) *The Dragons of Eden*, Coronet Books, London.

Schell, J. (1982) *The Fate of the Earth*, Picador, London.

Schiller, F.C.S. (1917) 'Scientific discovery and logical proof', in C. Singer (ed.) *Studies in the History and Method of Science*, vol. 1, Oxford University Press.

Schiller, F.C.S. (1921) 'Hypotheses', in C. Singer (ed.) *Studies in the History and Method of Science*, vol. 2, Oxford University Press.

Schilpp. P.A. (ed.) (1970) *Albert Einstein: Philosopher-Scientist*, Open Court, La Salle, Ill.

Schumacher, E.F. (1973) *Small is Beautiful*, Blond and Briggs, London.

Scientific American (1980) whole issue, 243 (3).

Seeley, R. (1948) *The Function of the University*, Oxford University Press.

Shotter, J. (1976) *Images of Man*, Methuen, London.

Silk, J. (1980) *The Big Bang*, Freeman, San Francisco.

SIPRI (1979) *Armament or Disarmament? The Crucial Choice*, Stockholm.

Skinner, B.F. (1973) *Beyond Freedom and Dignity*, Penguin, Harmondsworth.

Skolinowski, H. (1975) 'Knowledge and values', *The Ecologist*, 5 (1), pp. 8–15.

Smart, J.J.C. (1963) *Philosophy and Scientific Realism*, Routledge & Kegan Paul, London.

Snow, C.P. (1964) *The Two Cultures and a Second Look*, Cambridge University Press.

Sober, E. (1975) *Simplicity*, Clarendon Press, Oxford.

Swinburne, R. (ed.) (1974) *The Justification of Induction*, Oxford University Press.

Szasz, T. (1961) *The Myth of Mental Illness*, Hoeber-Harper, New York.

Teich, M. and Young, R.M. (eds) (1972) *Changing Perspectives in the History of Science*, Heinemann, London.

Thornley, J. (1981) *Workers' Co-operatives*, Heinemann, London.

Tichý, P. (1974) 'On Popper's definition of verisimilitude', *British Journal for the Philosophy of Science*, 25, pp. 155–60.

Townsend, P. (1979) *Poverty in the United Kingdom*, Penguin, Harmondsworth.

Truscott, B. (1943) *Red Brick University*, Faber, London.

Ulam, S.M. (1976) *Adventures of a Mathematician*, Scribner, New York.

UNESCO (1981a) *World Directory of Research Projects, Studies and Courses in Science and Technology Policy*, Paris.

UNESCO (1981b) *Research and Human Needs*, Paris.

Venables, E. and P. (1972) 'The study of higher education in Britain', in H.J. Butcher and E. Rudd (eds), *Contemporary Problems in Higher Education*, McGraw-Hill, New York, pp. 17–34.

Ward, B. (1972) *What's Wrong with Economics?*, Macmillan, London.

Ward, B. (1979) *Progress for a Small Planet*, Penguin, Harmondsworth.

Weber, M. (1947) *The Theory of Social and Economic Organization*, William Hodge, London.

Weber, M. (1949) *The Methodology of the Social Sciences*, Free Press, Chicago.

Webster, C. (1975) *The Great Instauration*, Duckworth, London.

Weinberg, S. (1977) *The First Three Minutes*, André Deutsch, London.

Werskey, G. (1978) *The Visible College*, Allen Lane, London.

Westland, G. (1978) *Current Crisis of Psychology*, Heinemann, London.

Wigner, E.P. (1970) *Symmetries and Reflections*, MIT Press.

Wilson, B.R. (ed.) (1974) *Rationality*, Blackwell, Oxford.

Wilson, E. (1975) *Sociobiology: the New Synthesis*, Harvard University Press.

Winch, P. (1958) *The Idea of a Social Science and its Relations to Philosophy*, Routledge & Kegan Paul, London.

Wootton, B. (1950) *Testament for Social Science*, Allen and Unwin, London.

Worsley, P. *et al*, (1970) *Introducing Sociology*, Penguin, Harmondsworth.

Worswick, G. (1972) *Uses of Economics*, Blackwell, Oxford.

Zamyatin, Y. (1972) *We*, Penguin, Harmondsworth.

Ziman, J. (1968) *Public Knowledge: the Social Dimension of Science*, Cambridge University Press.

Ziman, J. (1978) *Reliable Knowledge*, Cambridge University Press.

Zuckerman, S. (1982) *Nuclear Illusion and Reality*, Collins, London.

Index of Names